Table of Contents

Executive Summary 1

    INTRODUCTION..2

    SIGNIFICANCE OF DSRC 3

    STATUS OF DSRC..4

    ENVISIONED IMPLEMENTATION PATH 4

    KNOWN AND POTENTIAL GAPS 6

    OPPORTUNITIES FOR OTHER COMMUNICATIONS TECHNOLOGIES 8

    CONCLUSIONS 9

    ASSESSMENT FOR CONGRESS 11

Chapter 1 Introduction 12

Chapter 2 Significance of DSRC to Transportation..16

    2.I.   DSRC OVERVIEW 16

    2.II. USES OF THE 5.9 GHZ DSRC SPECTRUM...17

    2.III. ENTITIES INVOLVED IN DSRC 21

    2.IV. SIGNIFICANCE OF THE 5.9 GHZ DSRC BAND 22

    2.V. BENEFITS OF CONNECTED VEHICLE APPLICATIONS TO THE NATION 34

Chapter 3 Status of DSRC through Research and Development.......................38

    3.I.   OVERVIEW OF DSRC ENABLING TECHNOLOGIES 38

        3.I.A.   ESSENTIAL VEHICLE-BASED COMPONENTS 39

        3.I.B.   ESSENTIAL INFRASTRUCTURE COMPONENTS 41

    3.II. OVERVIEW OF THE DSRC CRASH AVOIDANCE SAFETY APPLICATIONS 43

    3.III. DSRC AND V2V COMMUNICATIONS SECURITY 45

        3.III.A. SECURITY CREDENTIAL MANAGEMENT SYSTEM (SCMS) 46

        3.III.B. SYSTEM INTEGRITY AND MANAGEMENT..48

        3.III.C. ORGANIZATION AND OWNERSHIP 49

        3.III.D. SYSTEM GOVERNANCE 51

    3.IV. ESSENTIAL STANDARDS...54

    3.V. DSRC MATURITY AND STATUS 55

        3.V.A. STATUS AND MATURITY OF THE DSRC CONNECTED VEHICLE TECHNOLOGIES AND APPLICATIONS 55

        3.V.B. STATUS AND MATURITY OF DSRC IMPLEMENTATION.........................59

        3.V.C. ANALYSIS OF KNOWN AND POTENTIAL GAPS 61

        3.V.D GAO REPORT TO CONGRESS ASSESSMENT ON V2V AND DSRC READINESS 68

3.V.E   RELATIONSHIP OF DSRC TO EMERGING AUTOMATED VEHICLE TECHNOLOGIES 69

Chapter 4 DSRC Path for Implementation 70

4.I.   IMPLEMENTATION OF CONNECTED VEHICLE ENVIRONMENTS 70

4.II.   ENABLING IMPLEMENTATION 75

     4.II.A.   FUNDING 75

     4.II.B.   LICENSING.. 76

     4.II.C.   ROLE OF THE NATIONAL ITS ARCHITECTURE 76

     4.II.D.   ROLE OF ITS STANDARDS 77

     4.II.E.   CERTIFICATION 80

Chapter 5   Conclusion 83

References   85

Appendix A. Glossary of Acronyms 87

Appendix B. Summary of Analyses of Communications Media Options 93

B.I.   2014: COMPARISON OF COMMUNICATIONS FEATURES FOR MEETING V2V CRASH-AVOIDANCE REQUIREMENTS 93

B.II.   2012-2014: INDEPENDENT ANALYSES ON COMMUNICATIONS TECHNOLOGIES 98

     B.II.A.   2012-2013: FOR SECURITY CREDENTIAL MANAGEMENT BY BOOZ ALLEN HAMILTON. 98

     B.II.B.   2013-2014: COMMUNICATIONS ANALYSIS FOR V2I SERVICES BY AASHTO 101

     B.II.C.   2013-2014: FOR BACKHAUL SERVICES AND APPLICATIONS BY AASHTO 106

Appendix C. DSRC in Use Today 108

C.I.   RESEARCH TEST BEDS USING DSRC 108

     C.I.A.   CALIFORNIA.. 108

     C.I.B.   MICHIGAN.. 109

     C.I.C.   NEW YORK STATE.. 109

     C.I.D.   ORLANDO, FLORIDA.. 110

     C.I.E.   ARIZONA 111

     C.I.F.   ANN ARBOR, MICHIGAN 112

     C.I.G.   NORTHERN VIRGINIA 113

     C.I.H.   CANADA 114

     C.I.I.   AFFILIATED CONNECTED VEHICLE TEST BEDS 116

C.II.   OPERATIONAL USES OF DSRC 116

     C.II.A.   SOUTHEASTERN MICHIGAN 2014 PROJECT.. 116

     C.II.B.   SAN FRANCISCO MUNICIPAL TRANSIT AUTHORITY....................................... 117

     C.II.C.   SEATTLE/KING COUNTY 117

C.III.   PLANNED USES FOR DSRC 118

     C.III.A.   CONNECTED VEHICLE PILOT DEPLOYMENTS 118

     C.III.B.   EUROPEAN UNION: ROTTERDAM TO VIENNA CORRIDOR 119

Appendix D. History of DSRC: Policy and Technical **120**

    D.I.   DSRC: POLICY HISTORY 121

        D.I.A.   1998-1999: HISTORY OF THE FCC RULE 121

        D.I.B.   2003-2006: HISTORY OF SPECTRUM SHARING WITH THE SATELLITE INDUSTRY 122

        D.I.C.   2012-PRESENT DSRC OPERATIONS 128

        D.I.D.   UPCOMING POLICY MILESTONES 131

    D.II.   DSRC: TECHNICAL HISTORY 137

        D.II.A   RESEARCH AND DEVELOPMENT OF DSRC TECHNOLOGIES AND APPLICATIONS........ 138

        D.II.B.   LESSONS LEARNED FROM TESTING 142

        D.II.C.   MATURING THE TECHNOLOGIES... 146

        D.II.D.   TECHNICAL MILESTONES... 150

Appendix E. Definition of Connected Vehicle Enabling Technologies................................... **153**

    E.I.   OVERVIEW OF THE ENABLING TECHNOLOGIES 153

        E.I.A.   IN-VEHICLE COMPONENTS 153

        E.I.B.   EXTERNAL V2V AND V2I COMPONENTS... 156

    E.II. STANDARDS 158

Appendix F. Definition of Connected Vehicle Safety Applications–V2V and V2I **161**

    F.I   V2V SAFETY APPLICATIONS...161

    F.II. V2I SAFETY APPLICATIONS 162

Appendix G. Proposed Certification Path **164**

Appendix H. International Uses of DSRC **167**

    H.I.   DIFFERENCES BETWEEN THE US REGIONAL VISION AND OTHER REGIONS 167

        H.I.A.   COMPARISON OF US TO EUROPEAN UNION (EU).................................... 167

        H.I.B   COMPARISON OF US TO ASIA 168

        H.I.C   AUSTRALIA.. 169

    H.II. HARMONIZATION OF INTERNATIONAL STANDARDS AND ARCHITECTURE
        AROUND THE VEHICLE PLATFORM 169

Appendix I. Spectrum Operations **171**

        I.I.A   CHANNEL SWITCHING MODE.. 172

        I.I.B.   MULTI-CHANNEL OPERATION VERSUS A DEDICATED SAFETY CHANNEL 175

        I.I.C.   INTEROPERABILITY PERFORMANCE REQUIREMENTS 175

Appendix J   National Research Council Comments..**177**

List of Tables

Table 2-1.   Safety Applications 20

Table B-1.   Comparison of Two-way Communications Technologies with Selected V2V Requirements 95

Table B-2.   Citations 96

Table E-1.   Cooperative System Standards for V2V and V2I Communications...............158

Table F-1.   Description of the V2V Safety Applications 161

Table F-2.   V2I Safety Applications 162

List of Figures

Figure 2-1.   Example of V2V Communications and a Warning Scenario 19

Figure 3-1.   Illustration of DSRC Vehicle-Based Technologies..40

Figure 3-2.   Basic External V2V and V2I Roadside Infrastructure Components for a Typical Deployment 43

Figure 4-1.   Candidate Paths to Deployment 74

Figure C-1.   Test Bed Sites 132

Figure D-1.   Safety Pilot Model Deployment – Heavy Vehicles 132

Figure D-2.   Illustration of the Vehicles & Devices in Safety Pilot 146

Figure E-1.   Illustration of DSRC Vehicle-Based Technologies..154

Figure E-2.   Basic External V2V and V2I Roadside Infrastructure Components for a Typical Deployment (Image Source: USDOT) 157

Figure I-1.   DSRC Channel Assignment 173

Figure I-2.   Time-Division Channel Usage..173

Figure I-3.   I-495 and I-66 Interchange 175

Executive Summary

On July 6, 2012, President Obama signed into law a two-year transportation reauthorization bill, the Moving Ahead for Progress in the 21st Century Act (MAP-21) (P.L. 112-141). This report responds to Section 53006, codified at 23 U.S.C. § 518:

"§ 518. Vehicle-to-vehicle and vehicle-to-infrastructure communications systems deployment

"(a) IN GENERAL.—Not later than 3 years after the date of enactment of this section, the Secretary shall submit to the Committees on Commerce, Science, and Transportation and Environment and Public Works of the Senate and the Committees on Transportation and Infrastructure, Energy and Commerce, and Science, Space, and Technology of the House of Representatives a report that—
   "(1) assesses the status of dedicated short-range communications technology and applications developed through research and development;
   "(2) analyzes the known and potential gaps in short-range communications technology and applications;
   "(3) defines a recommended implementation path for dedicated short-range communications technology and applications that—
    "(A) is based on the assessment described in paragraph (1); and
    "(B) takes into account the analysis described in paragraph (2);
   "(4) includes guidance on the relationship of the proposed deployment of dedicated short-range communications to the National ITS[1] Architecture and ITS Standards; and
   "(5) ensures competition by not preferencing the use of any particular frequency for vehicle to infrastructure operations.

"(b) REPORT REVIEW.—The Secretary shall enter into agreements with the National Research Council and an independent third party with subject matter expertise for the review of the report described in subsection (a)."

This report responds to a Congressional request for an assessment of the 5.9 Gigahertz (GHz) Dedicated Short Range Communications (DSRC) technology. A draft version of this report was provided for review and comment to a set of subject matter experts convened by the National Research Council (NRC) of the National Academies of Sciences, Engineering, and Medicine,

---

[1] Intelligent Transportation Systems

who further engaged with independent third parties to validate conclusions. The NRC Letter Report was delivered to Congress in April 2015 and is included in this report as Appendix J. This report has been revised as needed to respond to each technical comment in the NRC Letter Report.

Introduction

5.9 GHz DSRC is a Wi-Fi derivative technology developed to meet specialized needs for secure, low latency,[2] wireless mobile data communications. It is uniquely configured to support safety-critical applications through continuous, high-speed, trusted and authenticable wireless data communications among moving vehicles and between vehicles and roadway infrastructure or mobile devices.

In 1999, the Federal Communications Commission (FCC) allocated 75 Megahertz (MHz) of wireless spectrum at 5.850-5.925 GHz for Intelligent Transportation Systems (ITS) services using DSRC. This range of the spectrum is referred to as the 5.9 GHz band. Since then, the United States Department of Transportation (USDOT) has worked diligently and collaboratively with industry and public sector stakeholders to develop and evaluate new cooperative technologies, equipment, and applications known as Connected Vehicle (CV) technologies, inclusive of vehicle-to-vehicle (V2V), vehicle-to-infrastructure (V2I), and vehicle-to-devices or other points of connection (V2X).

At this time, no other wireless technology has proven the ability to provide all of the critical attributes of DSRC needed to support V2V and V2I safety applications. Whereas commercial wireless communications technologies continue to improve in terms of latency and security, none match DSRC performance capabilities or provide comparable user privacy and message authentication controls.

USDOT and stakeholder analysis continues to conclude that DSRC is the only viable option for safety-critical and other low latency mobility and environmental applications. The transmission of the mandated Basic Safety Message (BSM) by each connected vehicle requires the attributes afforded by DSRC in order to support crash-avoidance applications and produce safety-critical warnings to the driver in time for action to be taken. These same research efforts, however, demonstrate opportunities to employ non-DSRC communications technologies in support of connected applications that do not require the low latency, high reliability and availability of the 5.9 GHz band, or significant privacy protection. Examples of these latter applications include: mobility and logistics; environmental performance; road weather and traveler information;

---

[2] Latency is a measure of the time delay experienced in a system, usually between the sending, and subsequent reception, of information. The lower the latency, the faster the transmission. A more detailed definition is provided in Chapter II and highlights additional factors in the transmission of messages.

---

security credential management; field equipment-to-center communications (backhaul); or back-office operating agency communications and decision-support systems, among other uses.

Significance of DSRC

5.9 GHz DSRC is important to the Nation because it can be configured to enable real-time crash-avoidance alerts and warnings—offering a significant opportunity to achieve a transformation in transportation safety. DSRC permits data on emerging roadway hazards to be gathered from multiple external sources (i.e., other vehicles that are broadcasting data, infrastructure, or portable devices) and fused with on-board vehicle data. Because of the dedicated nature of the spectrum band and the low latency configuration of the communications technologies, safety-critical alerts and warnings can be provided to drivers in time to avoid a crash.

The National Highway Traffic Safety Administration (NHTSA) estimates that an initial set of DSRC-based V2V and V2I safety applications have the potential to address 83 percent of light-vehicle crashes involving unimpaired drivers.[3] NHTSA's analysis and field demonstration results have identified that a set of V2V applications offer unique capabilities to prevent crash-types which are not preventable through existing or emerging vehicle-based sensors and vehicle automation technologies alone. These applications are: [4]

- The Intersection Movement Assist application that addresses safety by warning drivers if an unsafe situation develops as vehicles approach and move through intersections;

- The Left-Turn Assist application that alerts drivers to hazards of on-coming traffic when making left turns; and

- The Emergency Electronic Brake Light application that enables a vehicle to warn its driver to brake in a situation where another V2V-equipped vehicle decelerates quickly but may not be directly in front of the warning vehicle. The EEBL warning is particularly useful when the driver's line of sight is obstructed by other vehicles or bad weather conditions, such as fog or heavy rain.

---

[3] Results from *Frequency of Target Crashes for IntelliDrive Safety Systems*, Najm, W., J. Koopman, S. Smith, and J. Brewer, October 2010, DOT HS 811 381. See: http://www.nhtsa.gov/Research/Crash+Avoidance/ci.Office+of+Crash+Avoidance+Research+Technical+Publications.print (last accessed January 30, 2014).

[4] The detailed analyses on which effectiveness estimates of these critical safety applications can be found in the report: *Vehicle-to-Vehicle Communications: Readiness of V2V Technology for Application* which is available in the public docket (http://www.nhtsa.gov/staticfiles/rulemaking/pdf/V2V/Readiness-of-V2V-Technology-for-Application-812014.pdf. Additional reports related to the independent evaluation of the safety pilot model deployment and a comprehensive safety benefits analysis for all applications testing during the model deployment will be released to the public in the coming months.

Status of DSRC

DSRC is sufficiently robust to proceed with preparations for deployment of Connected Vehicle environments.

In 2012, research on Connected Vehicle technologies culminated in a large-scale model deployment and evaluation of essential, prototype DSRC technologies and V2V and V2I safety applications. The Safety Pilot model deployment was performed with drivers in over 2,800 vehicles (cars, buses, trucks and motorcycles) and on a range of device types designed and produced by multiple manufacturers. Results indicated that safety applications using V2V technology are interoperable and can address a large majority of crashes involving two or more motor vehicles. These results supported NHTSA's decision to move forward with a V2V communications rulemaking for light vehicles. NHTSA announced their decision in an Advanced Notice of Proposed Rulemaking (ANPRM) in August 2014, and documented their foundational analysis in an accompanying report titled, "*Vehicle-to-Vehicle Communications: Readiness of V2V Technology for Applications*" (herein referred to as the NHTSA Technology Readiness report).[5]

Findings from the Safety Pilot model deployment and related research have resulted in actions by industry suppliers to improve DSRC-based applications, equipment, and technologies. Evaluation of the standards validated the ability to support interoperability needed across different vendor DSRC devices. Results further highlighted areas for refinement, noting that, in many cases, DSRC is viable but that clarity on how to implement the associated standards is needed, indicating the necessity for equipment certification processes. Since the publication of early results in 2013, vendors and standards working groups have addressed issues as a means of ensuring interoperability. USDOT has also engaged with industry to establish certification test processes and procedures to be used for entities participating in certification testing.[6]

Envisioned Implementation Path

Working with stakeholders, USDOT has synthesized analyses and preliminary plans into a roadmap to guide industry and public agency implementation efforts. The roadmap anticipates advancements in implementation during three overlapping timeframes—2015-2025, 2020-2035, 2030-2040—during

---

[5] NHTSA ANPRM [Docket No. NHTSA-2014-0022]: http://www.nhtsa.gov/About+NHTSA/Press+Releases/NHTSA-issues-advanced-notice-of-proposed-rulemaking-on-V2V-communications.

[6] There are two certification paths, both of which provide the ability to obtain security credentials. The first path, self-certification, applies to automotive manufacturers who will comply with NHTSA requirements to include DSRC capabilities in their vehicle. The second path is for device manufacturers or application developers to have their products certified through certification laboratories that are now being established to provide such services.

---

which the preponderance of communications capabilities within the vehicle fleet will evolve. It offers three milestones for the year 2040[7]:

1. 90 percent of the US light duty vehicle fleet is <u>DSRC enabled</u>;

2. 80 percent of all traffic signals are <u>DSRC equipped</u>; and

3. DSRC exists at 25,000 <u>additional safety-critical</u> roadway locations.

The roadmap is predicated primarily on a market-driven scenario wherein applications are provided by industry for consumer adoption and public agency implementation. Along with selective infrastructure investments, it is expected that public agencies will also establish public-private partnerships that result in Connected Vehicle operations that meet regional and local priorities and that further allow for innovative market opportunities in the areas of mobility and environmental services.

The roadmap assumes an austere near-term budgetary environment for both public and private sectors, and builds upon existing and emerging DSRC footholds established through test and operational environments already in seven states, and others anticipated to be developed in conjunction with upcoming regional pilot deployments.[8]

The roadmap also anticipates continued advancements in sensor-based vehicle crash avoidance and automation technologies, which will leverage, rather than supplant, V2V communications capabilities. DSRC applications are advancing in a similar manner in the European Union (EU), Japan, South Korea, Australia, and Canada. USDOT is working to harmonize operational policies and voluntary industry standards to achieve global compatibility and to facilitate domestic access to international markets.

The National ITS Architecture and ITS Standards accommodate both DSRC and other wireless and wireline communication technologies. A Connected Vehicle Reference Implementation Architecture (CVRIA) has been prepared to illustrate various approaches to implement Connected Vehicle applications. The CVRIA offers choices for using different technologies, and highlights the use of key technical standards.

---

[7] *National Connected Vehicle Field Infrastructure Footprint Analysis - Final Report (FHWA-JPO-14-125),* AASHTO/USDOT; June 27, 2014 at
http://stsmo.transportation.org/Documents/AASHTO%20Final%20Report%20_v1.1.pdf (Last accessed June 2015).
Embedded cellular data communications capabilities in the light duty fleet are also anticipated by this time.
[8] For more information on the seven states and two Canadian sites, see: http://www.its.dot.gov/testbed.htm and Appendix C. For more information on the upcoming pilot deployments, see: http://www.its.dot.gov/pilots/index.htm.

Known and Potential Gaps

MAP-21 Section 53006 (a) requires, in part, that the DSRC report analyze known and potential gaps in short-range communications technology. DSRC technology is sufficiently mature that gaps predominantly have been addressed. What remains is the work to provide greater specificity on issues such as spectrum usage, performance requirements, and final standards to ensure the interoperability of devices.

Recently, two issues have arisen that are providing uncertainty for vendors making investments and early adopters formulating plans.

From a technical perspective, the first issue is the proposal to allow unlicensed Wi-Fi devices to operate in the DSRC band. As implementers of Connected Vehicle technologies make investment decisions, they require the assurance that prospective sharing of the spectrum band dedicated for V2V and V2I communications will not jeopardize DSRC-based crash avoidance capabilities. Unlicensed wireless broadband devices (referred to as Unlicensed-National Information Infrastructure or U-NII devices) are expected to become widespread in the foreseeable future. It is important to ensure that changes to the 5.9 GHz DSRC band do not jeopardize safety-critical crash avoidance capabilities. U-NII devices have yet to be tested to determine if they will interfere with crash-avoidance applications or result in unacceptable risks to traveler safety.

Second, from an organizational perspective, the emerging issue is the possibility that interference with the 5.9 GHz band could negatively impact the delivery of communications that form the basis for safety-critical warnings to drivers. If this were to happen, active frequency coordination might be required to address the problem.

In addition to these two known issues that could impact successful deployment, there exists a small set of open items that need further specificity to support implementers. Many of these, such as liability clarification or driver-vehicle interface requirements, were identified as gaps by the NRC, and are addressed in the NHTSA Technology Readiness report that accompanied the V2V ANPRM. Section 3.V.C provides a more detailed discussion of these items.

Unlicensed Users of 5.9 GHz Band

With regard to whether the DSRC safety applications can co-exist with unlicensed wireless broadband devices, USDOT is engaged with industry-based working groups to learn about the device properties. USDOT is working with experts and stakeholders to develop test plans and facilities for evaluating the potential for harmful interference from U-NII devices with the intent of ensuring interference-free operations of crash-avoidance safety systems. USDOT has begun testing and is actively seeking industry partners to provide unlicensed devices for testing with DSRC devices. At this time, it is anticipated that results will be available - by December of 2016 (dependent upon various factors, including when prototype U-NII devices are provided). These test results will inform the FCC and National Telecommunications and Information Administration (NTIA) and address their 2013 report that evaluated spectrum-sharing technologies and the risk to users if U-NII devices were allowed to operate in the 5.9 GHz DSRC band.[9] Section 3.V.C of this report provides a discussion of USDOT's approach to addressing the spectrum sharing issue.

Frequency Coordination

As DSRC technology has developed and the rules governing its use have solidified, stakeholders and experts are highlighting a potential institutional gap associated with a desire for a more active form of frequency coordination. Two underlying concerns are:

- The potential for interference amongst proximate, unrelated DSRC roadside units (RSUs); and

- The potential need to address interference associated with use of unlicensed devices. Federal Aviation Administration (FAA) experience with the Terminal Doppler Weather Radar (TDWR) systems provides a reference point for this potential need.

To mitigate these concerns, technical spectrum coordination techniques and organizational management roles have been under discussion. Although the FCC declined to establish such an approach when requested in 2004,[10] the requirements for using spectrum associated with safety-critical applications are now better understood.

USDOT is performing research to define the needs associated with frequency management, and to clearly and concisely propose the coordination that would be needed with FCC and NTIA.

More work is needed to clearly and concisely describe the frequency coordination issue, its potential impact on safety, and its possible effects on DSRC deployments around the country if not effectively addressed. USDOT is working with stakeholders, including the FCC and NTIA, to identify characteristics needed for entities to become spectrum managers, including financial sustainability, as well as to propose roles.

---

[9] NTIA January 2013 report titled *Evaluation of the 5350-5470 MHZ and 5850-5925 MHZ Bands Pursuant to Section 6406(b) of the Middle Class Tax Relief and Job Creation Act of 2012*. Located at: http://www.ntia.doc.gov/files/ntia/publications/ntia_5_ghz_report_01-25-2013.pdf.

[10] See Amendment of the Commission's Rules Regarding Dedicated Short-Range Communication Services, WT Docket No. 01-90 at: http://apps.fcc.gov/ecfs/document/view;jsessionid=sK3cQdpGLCGPDp1bQCdybsRVbYjkqbc1lLDjkL9yn93XlypvyLhQ!973241960!-856245186?id=6516483012

Opportunities for Other Communications Technologies

In MAP-21, Congress stated a desire to learn how V2I implementations will not preference the use of any particular communications technology, specifically for V2I operations. Over the years, USDOT has regularly performed comparative analyses of communications options to ensure that multiple choices would be available for use within Connected Vehicle environments. While analyses continue to conclude that DSRC is, to date, the only viable option for safety-critical applications, these same research efforts demonstrate opportunities to use other, commercially-available, wireless data communications technologies (such as cellular, satellite, radio, fiber, and Wi-Fi, among others) in support of applications that do not require the extreme low latency and fast network access (connection) times afforded by DSRC. Examples of these applications include, among others:

- Mobility and logistics
- Environmental performance
- Road weather and traveler information
- Security credential management
- Field equipment-to-center (backhaul) communications
- Agency communications or decision support systems.

These other communications media can be employed for implementations today and are expected to be fully leveraged in the upcoming CV Pilot sites.[11] Additionally, new communications technologies are in development, such as 5G or LTE (Long-Term Evolution) Direct that may also have advantages for various V2I application categories. USDOT and the ITS Standards community are monitoring and considering these developments to determine whether and how these might be employed effectively for Connected Vehicle applications. Importantly, incorporating these other types of communications requires additional advancements to support seamless data exchange and connectivity among these and DSRC. USDOT has launched development of new message protocols and security credential management solutions that will allow seamless exchange across a range of wireless data communication technologies.

New requirements also may need to be addressed in the following areas for these technologies to be used in a cooperative environment:

- **Security**—non-DSRC communications will need to provide security credentials in the same or a compatible manner with Connected Vehicle security to ensure that messages will be authenticated and trusted. A range of practices exists in today's communications

---

[11] For more information on the CV Pilots, please see: http://www.its.dot.gov/pilots/index.htm. (Last accessed June 2015)

services, from no security at all to proprietary security. Consumers opt-in to using applications that allow data transfers based on their acceptance of the form and level of security. Connected Vehicle safety applications, in particular, require a higher level of security and a more rigorous certification process than today's devices and applications used in commercial domains where life safety is not at stake.

- **Reliability**—non-DSRC communications providers will need to demonstrate how they can provide reliable and timely data messaging, especially under congested conditions. For example, mobility applications related to vehicle emissions reductions and fuel consumption reductions will need frequent communications with traffic signals in order to provide drivers actionable feedback on how to time their approach and departure at intersections. A communications delay of several seconds or momentary unavailability that may be possible with cell phone-based communication under highly congested conditions would disrupt or disable such applications.

- **Privacy**—the mandated DSRC technologies incorporate technical controls to appropriately mitigate potential consumer privacy risk. In contrast, the majority of V2I and V2X applications that use non-DSRC communications are expected to be opt-in, and consumers may decide on the levels of privacy and personal data capture/data use that is acceptable to them. For the DSRC technologies, USDOT has worked with industry experts to develop a security solution that has high levels of privacy protection designed into it—the data exchange is still effective despite not knowing information about the user.

Conclusions

DSRC represents the most cost effective approach to realizing enormous traffic safety benefits, and offers the ability to enhance mobility and environmental vehicular applications. In summary, this report concludes the following:

- **5.9 GHz DSRC remains a foundational requirement for enabling safety-critical V2V and V2I applications.** Vital characteristics are:
  - Rapid transmission speed, low latency, stability, and dedicated availability make DSRC the optimal communication media currently available for safety-critical applications, particularly those focused on crash avoidance; and
  - DSRC offers the highest levels of privacy because USDOT has ensured that the mandated BSM received by roadside equipment (RSEs) from V2V-enabled vehicles will not personally identify the vehicle (as through a Vehicle Identification Number (VIN) or registration number) or the vehicle's driver or owner.

- **Operations that use DSRC—test beds, operational sites, and emerging pilot sites—are demonstrating how the spectrum is used**; specifically finding that:
  - The FCC's existing service rules and spectrum band plan for DSRC are viable.
  - Other forms of communications technologies–cellular, Wi-Fi, satellite, fiber, etc.— are viable for mobility, environmental, and other non-safety critical applications.

- **DSRC is ready for wider-scale implementation**:
  - USDOT has synthesized stakeholder deployment plans into a roadmap to illustrate a candidate path for implementation.
  - Vendors and standards working groups have used the results from the Safety Pilot Model Deployment tests to improve and refine their products to be production-ready and available for implementation in the near future.
  - USDOT is preparing the policy foundation:
    - NHTSA has announced its intention to complete a V2V Notice of Proposed Rulemaking (NPRM) by early 2016 with regard to DSRC-based technologies for light vehicles;
    - The Federal Highway Administration (FHWA) is issuing V2I Guidance for State and local agencies along with tools and reference guidelines; and
    - The ITS Joint Program Office (JPO) is working on a comprehensive standards plan; is preparing for integration of the Connected Vehicle Reference Implementation Architecture into the National ITS Architecture; has produced an architecture tool for State and local implementers; and is continuing to pursue opportunities to harmonize its architecture and standards on an international level.
  - An initial security credential management system (SCMS) will be available in 2016 to support early deployments.
  - In incentivizing the first wave of CV Pilot deployments in 2015, the USDOT seeks operational deployments of V2V, V2I, and V2X applications that synergistically capture and utilize new forms of Connected Vehicle and mobile device data to improve multimodal surface transportation system performance and enable enhanced performance-based systems management in an observable and measureable near-term manner.
- **With regard to calls for spectrum sharing—completion of analysis, testing, and simulation modeling in 2016/2017 will provide details necessary to further inform the FCC's exploration of sharing technologies on whether:**
  - Co-existence of Connected Vehicle technologies with unlicensed devices is proven feasible.
  - Proposed spectrum sharing plans or technologies can prove, conclusively and practically, the capability to avoid harmful interference with crash-avoidance safety systems. Upcoming tests will provide necessary insights to inform policy.

Throughout their ongoing research, USDOT and its partners have evaluated how the allocated DSRC spectrum will be applied. As of mid-2015, approximately 20 implementing agencies and some academic and private sector organizations have licenses that allow for use of DSRC for experimental installations. These licensed sites will form the basis for initial operations in those areas. Descriptions of these sites are provided in Appendix C of this report. They show the interest in advancing to this next generation of ITS by agencies in: Arizona, California (multiple sites including a transit operations site in San Francisco), Florida, Michigan (multiple sites), New York, Virginia (multiple sites), and Washington, and across the border in Canada (Alberta and British Columbia), where there are ongoing discussions about architecture harmonization.

Assessment for Congress

USDOT is confident that DSRC is ready for deployment and that DSRC-based technologies and applications offer a necessary and essential path to a safer and more efficient surface transportation system for America.

USDOT finds that there are no significant gaps in DSRC technologies or applications. Recent, emerging "unknown issues" include requests for coexistence with widespread use of broadband unlicensed devices, and a desire for spectrum management that provides an active response to problems as they arise.

A critical assumption in the development of DSRC-based technologies and applications was that spectrum sharing of this nature was not intended. To ensure that changes in the 5.9 GHz DSRC band do not jeopardize crash avoidance capabilities, USDOT is working with industry partners and in cooperation with the FCC and NTIA to define and test for harmful interference.

With the assessment that DSRC is ready for use, USDOT, industry partners, State and local agencies, and key stakeholders are defining paths for implementation. USDOT has developed a candidate roadmap that uses as its foundation:

- Stakeholder analyses that examine the local conditions for implementation;

- The candidate architecture views in the Connected Vehicle Reference Implementation Architecture; and

- An analysis of the standards needed to support DSRC operations as well as the more comprehensive needs for sustaining Connected Vehicle environments.

With its research efforts, USDOT has acted to ensure competition by not showing preference in the use of any particular frequency for non-safety related V2I operations. To support an implementer's ability to leverage a wide range of communications options, USDOT is advancing state-of-the-industry practices in security credential management and message protocols, data distribution and capture methodologies, and data warehousing.

Chapter 1 Introduction

The Moving Ahead for Progress in the 21st Century Act (P.L. 112-141; also known as MAP-21) requires the Secretary of the United States Department of Transportation (USDOT) to produce a report on the status of Dedicated Short-Range Communications (DSRC):

> *"(a) IN GENERAL.—Not later than 3 years after the date of enactment of this section, the Secretary shall submit to the Committees on Commerce, Science, and Transportation and Environment and Public Works of the Senate and the Committees on Transportation and Infrastructure, Energy and Commerce, and Science, Space, and Technology of the House of Representatives a report that—*
>> *"(1) assesses the status of dedicated short-range communications technology and applications developed through research and development;*
>> *"(2) analyzes the known and potential gaps in short-range communications technology and applications;*
>> *"(3) defines a recommended implementation path for dedicated short-range communications technology and applications that—*
>>> *"(A) is based on the assessment described in paragraph (1); and*
>>> *"(B) takes into account the analysis described in paragraph (2);*
>> *"(4) includes guidance on the relationship of the proposed deployment of dedicated short-range communications to the National ITS Architecture and ITS Standards; and*
>> *"(5) ensures competition by not preferencing the use of any particular frequency for vehicle to infrastructure operations.*
>
> *"(b) REPORT REVIEW.—The Secretary shall enter into agreements with the National Research Council and an independent third party with subject matter expertise for the review of the report described in subsection (a)."*

DSRC is the only wireless technology that provides trusted and secure, low latency,[12] wireless data communications configured to meet the unique needs of high-speed data exchange among moving vehicles and with roadway infrastructure devices without compromising personal privacy or facilitating the tracking of traveler whereabouts. These attributes are essential to safety-critical crash avoidance technologies and applications within a free society. This report

---

[12] Latency is a measure of the time delay experienced in a system, usually between the sending, and subsequent reception, of information. The lower the latency, the faster the transmission. A more detailed definition is provided in Chapter II and highlights additional factors in the transmission of messages.

documents the analysis that concludes DSRC is, to date, the only viable option for safety-critical applications.

These same research efforts, however, also illustrate opportunities to use other communications technologies in support of applications that do not require low latency, high reliability and availability of spectrum, or privacy. These applications are expected to include applications for (but not limited to):

- Mobility and logistics
- Environmental performance
- Road weather and traveler information
- Security credential management
- Field equipment-to-center (backhaul) communications
- Agency back-office communications.

This report provides an assessment of the status of DSRC technology and applications, including known and potential gaps; describes a recommended implementation path; and discusses opportunities to use commercially available communications for Connected Vehicle environments under certain circumstances. It does so in five chapters with supporting appendices. Over ten years of research, development, prototyping, and testing have generated a wide range of analyses, reports, requirements documents, and briefings that address DSRC from different perspectives. The appendices to this report gather the essential and foundational details in one place to provide a deeper background and context on DSRC.

Title	Contents
Chapter 1 Introduction	This chapter describes the purpose for this report and the organization of the report's contents.
Chapter 2 Significance of DSRC to Transportation	Chapter 2 describes: • The use of the 5.9 GHz DSRC band for Connected Vehicle technologies and applications; • The significance of the 5.9 GHz DSRC band's characteristics that make it uniquely suited to crash-avoidance safety; and a comparison of DSRC to other communications; and • An illustration of the benefits that are anticipated from having dedicated bandwidth in support of crash prevention.

Title	Contents
Chapter 3 **An Assessment of the Status of DSRC through Research and Development**	Chapter 3 provides an assessment of the status of: • DSRC technologies (both in-vehicle and infrastructure-based); • Crash avoidance applications; • Essential cooperative system standards; • DSRC and V2V communications security; • The implementation of DSRC; • The identification of essential standards • Conclusions on the maturity and status of DSRC, including discussion of known and potential gaps; • The Government Accountability Office's (GAO) assessment of V2V readiness; and • DSRC's relationship to Automated Vehicle technologies.
Chapter 4 **Recommended Path for DSRC Implementation**	Chapter 4 offers a vision for implementation, one that is based on analysis and real-world experiences and lessons learned; and is grounded in the National ITS Architecture and ITS Standards. This chapter also describes: • How USDOT has ensured the ability to utilize a wide range of communications media for implementing Connected Vehicle environments; and • A roadmap for finalizing research in support of implementation.
Chapter 5 **Conclusion**	Chapter 5 provides conclusions that address the Congressional requirements in MAP-21 sec. 53006.
References	This section provides a list of the references used to prepare this report.
Appendix A. Glossary of Acronyms	Appendix A offers a glossary of acronyms found throughout this report.
Appendix B. Summary of Analyses of Communications Media Options	Appendix B presents a summary of the analyses led by USDOT and performed by partners, experts, and stakeholders regarding the communications media options for Connected Vehicle environments.

Title	Contents
Appendix C. DSRC in Use Today	Appendix C lists and describes the many State and local areas (including emerging sites in Canada) that have deployed and are researching and/or operating Connected Vehicle environments; and are using DSRC and leveraging non-DSRC communications technologies.
Appendix D. History of DSRC Policy and Technical	Appendix D is a compilation of the significant policy milestones that provide the basis for transportation safety as a primary allocation in the band; and the relevant technical research and development milestones that have resulted in DSRC technologies and applications being available today. Appendix D also discusses the international strategies for using DSRC and the efforts to harmonize operational policies and standards to achieve global compatibility and facilitate domestic access to international markets.
Appendix E. Definition of Connected Vehicle Enabling Technologies	Appendix E presents definitions for the enabling technologies of a Connected Vehicle environment.
Appendix F. Definition of Connected Vehicle Safety Applications— V2V and V2I	Appendix F defines the key transformative safety applications in which USDOT, industry partners, academia, and stakeholders have invested.
Appendix G. Proposed Certification Path	Appendix G proposes a certification approach for ensuring that devices and applications meet performance requirements, utilize the DSRC band according to the rules of use, and are interoperable.
Appendix H. International Uses of DSRC	Appendix H describes how other nations are proposing to use the DSRC spectrum.
Appendix I. Spectrum Operations	Appendix I describes the spectrum allocation and operations for Connected Vehicle technologies and applications.
Appendix J. NRC Comments	Appendix J provides the NRC's technical comments in the form of a Letter Report.

Chapter 2 Significance of DSRC to Transportation

Chapter 2 provides an overview of the significance of DSRC to emerging and transformative Connected Vehicle technology and applications. The chapter is divided into three sections:

2.I.   Provides an overview of DSRC.

2.II.   Highlights the use of the 5.9 GHz DSRC band for crash-avoidance safety.

2.III. Describes the entities that are expected to participate in the commercialization and deployment of DSRC.

2.IV.   Discusses the significance of the 5.9 GHz DSRC band characteristics that make it uniquely suited to crash-avoidance safety; including a comparison of DSRC to other communications media.

2.V.   Illustrates the benefits that are anticipated from having a dedicated band for Connected Vehicle safety application purposes.

2.I. DSRC Overview

DSRC is a two-way wireless radio service that provides a high speed, reliable, short-range communication media for use among moving vehicles and between moving vehicles roadside infrastructure. DSRC is extremely efficient and robust because it permits all equipped vehicle to share critical safety information continuously. DSRC operates in constant broadcast-and-receive mode, providing situational information that is updated through repeated contact with other equipped vehicles in the vicinity. Each vehicle then independently evaluates the information received from nearby vehicles to determine if a warning message to the driver is required.

Each DSRC-equipped vehicle broadcasts a basic safety message (BSM) on the 5.9 GHz spectrum. The BSM is a small packet of data on vehicle situational elements, including:[13]

---

[13]As currently developed, the BSM will consist of 2 parts: Part 1 containing the core data elements needed for safety applications, and Part 2 containing supplemental data elements that will be broadcast only as needed, such as anti-lock brake system activation, or road-weather related data collected by the vehicle's sensors. For more information see: http://www.its.dot.gov/itspac/october2012/PDF/data_availability.pdf (Last accessed June 2015).

- Vehicle size
- Vehicle position (per global positioning system, or GPS)
- Speed
- Heading
- Steering angle
- Brake status

DSRC uses the 5.9 GHz spectrum efficiently, by broadcasting the BSM frequently (every 100 milliseconds). Other communication technologies are not suitable for these tasks because these have been designed for other purposes and are based on one-to-one, not one-to-many, communications. These other designs significantly reduce efficiency because of their need to establish one-to-one connections. This adds a significant overhead to the individual messages using the bandwidth which further reduces the timeliness of the information available by introducing delays into the process.

DSRC does not require "multiple-hops" between source and destination, including infrastructure, in the way cellular or common Wi-Fi communications do, nor does it encounter "network join" delays. Since the source and the recipient do not need to connect directly, privacy can be protected as long as a way to establish trust between each party is available (e.g., security credentials).

DSRC has been designed to work well in a rapidly moving environment, where the sender and a receiver may be moving toward or away from one another at speeds greater than 100 miles per hour.

Due to its unique combination of attributes (broadcast messaging that requires no network connection; messages small enough to be broadcast frequently and processed quickly; anonymous and trusted communications; and robust functionality in a rapidly moving environment), DSRC is, to date, the only viable option for safety critical vehicle-to-vehicle (V2V) communications.

2.II. Uses of the 5.9 GHz DSRC Spectrum

In the initial FCC Report and Order, the FCC allocated 75 MHz of spectrum in the 5.9 GHz DSRC band for use by Intelligent Transportation Systems (ITS) vehicle safety and mobility applications. The FCC noted the benefits of DSRC "...*to improve traveler safety, decrease traffic congestion, facilitate the reduction of air pollution, and help to conserve vital fossil fuels*."[14] New transportation safety technologies—known as Connected Vehicle technologies—are the primary focus of DSRC use.[15]

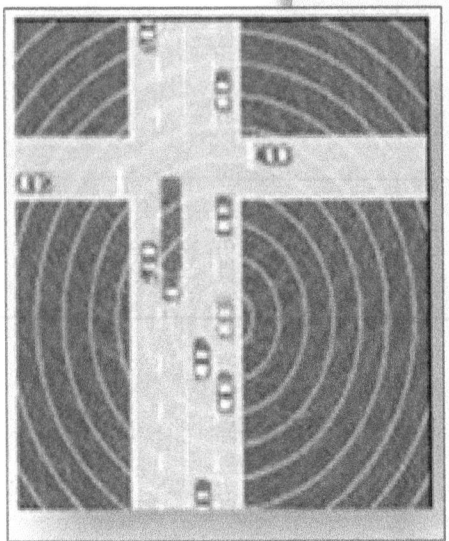

"Connected Vehicle Technologies"...

....are equipment, applications, systems, or technologies that enable vehicle-to-vehicle (V2V), vehicle-to-infrastructure (V2I), and vehicle-to-portable device (V2X) communications in support of communications-based safety, mobility, and environmental approaches to transportation. A key difference from today's traffic applications that connect travelers through smart phones or in-vehicle navigation systems, is the additional data that can be drawn from electronic control units and sensors connected to a vehicle's Controller Area Network bus (CAN bus). This data is broadcast to nearby vehicles, providing the receiving vehicle with 360 degrees of coverage and detection. Today's vehicle-based sensors are limited in terms of direction, field-of-view, and distance at which they are able to detect potential conflicts. Short-range data exchange with infrastructure can produce specific road-segment information for drivers and travelers within a specific geographic area. Image: USDOT

Connected Vehicle technologies, in particular V2V technologies, are expected to reduce the severity of or prevent motor vehicle crashes across the United States that result in numerous injuries and deaths and high economic costs.[16] According to NHTSA, if V2V technologies <u>alone</u> are widely deployed, they have the potential to address 81 percent[17] of light-vehicle crashes

---

[14] See FCC Report and Order 99-305, October 22, 1999 at: https://transition.fcc.gov/Bureaus/Engineering_Technology/Orders/1999/fcc99305.txt and FCC Report and Order 03-324, February 10, 2004, p.6 at: https://apps.fcc.gov/edocs_public/attachmatch/FCC-03-324A1.pdf. (Last accessed June 2015)

[15] Ibid.

[16] 5.3 million police-reported vehicle crashes occurred in the United States in 2011, resulting in about 32,000 fatalities and more than 2.2 million injuries, according to NHTSA statistics. The motor vehicle crashes caused $871 billion in economic loss and societal harm. This includes $277 billion in economic costs—nearly $900 for each person living in the United States based on calendar year 2010 data—and $594 billion in harm from the loss of life and the pain and decreased quality of life due to injuries.

[17] This percentage is consistent with the percentage on page 3 which further includes additional benefits when adding some of the V2I safety capabilities.

---

involving unimpaired drivers by providing warnings to drivers about emerging, real-time threats and hazards.[18]

Figure 2-1, below, illustrates how a vehicle equipped with DSRC technologies and V2V safety applications is able to analyze its own internal data (such as vehicle speed and location) and process data broadcast from other vehicles to assess emerging threats and hazards within the roadway. When on-board V2V applications sense that a collision with another similarly equipped vehicle is imminent, they can issue a warning to the vehicle driver.

Source: GAO.

Figure 2-1. Example of V2V Communications and a Warning Scenario

In this scenario, the truck and sports utility vehicle are at risk of colliding because the drivers are unable to see one another approaching the intersection and the stop sign is not visible to the driver of the truck because it has been damaged. Both drivers would receive warnings of a potential collision, allowing them to take actions to avoid it.

Table 2-1 identifies additional DSRC-based crash avoidance safety applications enabled by the availability of V2I and V2V communications. DSRC provides a 300-meter detection range, which represents a significant improvement over existing state-of-the-art sensor-based technologies (e.g., camera or radar systems), which can reach just 150 meters. Furthermore, due to the

---

[18] See original analysis at:
http://www.nhtsa.gov/DOT/NHTSA/NVS/Crash%20Avoidance/Technical%20Publications/2010/811381.pdf.

sharing of data between vehicles, V2V communications can provide the vehicle and driver with un-obstructed 360-degree situational awareness, resulting in driver alerts and warnings about potential collisions that are not visible to existing sensor-based technologies.

Table 2-1. Safety Applications[19]

V2I Safety	V2V Crash Avoidance Safety
• Red Light Violation Warning • Curve Speed Warning • Stop Sign Gap Assist • Spot Weather Impact Warning • Reduced Speed/ Work Zone Warning • Pedestrian in Signalized Crosswalk Warning (Transit)	• Emergency Electronic Brake Lights (EEBL) • Forward Collision Warning (FCW) • Intersection Movement Assist (IMA) • Left Turn Assist (LTA) • Blind Spot/ Lane Change Warning (BSW/LCW) • Do Not Pass Warning (DNPW) • Vehicle Turning Right in Front of Bus Warning (Transit)

From 2012-2014, USDOT and partners performed a pilot test of DSRC safety applications under real-world driving conditions. The pilot test results validated the deployment feasibility of V2V technology. Results, along with laboratory simulation, allowed NHTSA to estimate the preliminary safety benefits and initiate a rulemaking on V2V technologies on February 3, 2014.

The following is a brief summary from the February 2014 NHTSA announcement, "*U.S. Department of Transportation Announces Decision to Move Forward with Vehicle-to-Vehicle Communication Technology for Light Vehicles*":[20]

> *DOT research indicates that safety applications using V2V technology can address a large majority of crashes involving two or more motor vehicles. With safety data such as speed and location flowing from nearby vehicles, vehicles can identify risks and provide drivers with warnings to avoid other vehicles in common crash types such as rear-end, lane change, and intersection crashes. These safety applications have been demonstrated with everyday drivers under both real world and controlled test conditions.*

---

[19] The full list can be found at: http://www.its.dot.gov/pilots/cv_pilot_plan.htm
[20] Announcement from February 3, 2014 can be found here:
http://www.nhtsa.gov/About+NHTSA/Press+Releases/2014/USDOT+to+Move+Forward+with+Vehicle-to-Vehicle+Communication+Technology+for+Light+Vehicles. (last accessed June 1, 2015).

USDOT's commitment to these applications highlights current positions on safety:

- Safety is the highest priority for USDOT and forms the central focus for the research in support of a connected environment for transportation.

- Analysis illustrates that, to date, DSRC is the only available technology in the near-term that offers the latency, accuracy, and reliability needed for crash avoidance safety.[21]

2.III. Entities Involved in DSRC

A wide variety of entities have participated in the design, development, and testing of DSRC to date; additional entities are expected to participate in the commercialization and deployment of DSRC. Key participants include:

- **USDOT** directed and led early-stage DSRC development and testing. USDOT's ITS JPO coordinated a test bed in southeast Michigan involving 2,800 connected vehicles. USDOT is now enhancing and building the next generation of test beds in Florida, Northern California, and New York.[22]

- **FCC** develops frameworks and rules governing radio-spectrum resources. FCC licenses DSRC technologies that are fixed along roadsides to government entities, such as State, city, and county governments, and to public safety and business entities that meet eligibility requirements.[23] These fixed DSRC technologies make up the infrastructure part of V2I. Vehicle based devices are licensed by rule under Part 95 Subpart L[24] of the FCC's rules and regulations.

- **NTIA**, which is housed in the Department of Commerce, manages federal use of spectrum and advises the President on telecommunications policy issues. The Institute for Telecommunication Sciences located in Boulder, Colorado (**ITS Boulder**), is the research and engineering arm of NTIA.[25]

- **NHTSA** creates rules regarding connected vehicles. NHTSA is currently deciding whether to require basic DSRC capabilities in new cars.

---

[21] *Transforming Transportation through Connectivity: ITS Strategic Research Plan, 2010–2014: Progress Update, 2012.*

[22] ITS. http://www.its.dot.gov/testbed/connected_vehicle_testbeds_progress.htm (Last accessed June 2015)

[23] FCC. http://wireless.fcc.gov/services/index.htm?job=service_home&id=dedicated_src (Last accessed June 2015)

[24] FCC Part 95, Subpart L – Dedicated Short-Range Communications Service On-Board Units (DSRCS-OBUs) (see http://www.gpo.gov/fdsys/pkg/CFR-2009-title47-vol5/pdf/CFR-2009-title47-vol5-part95.pdf) (last accessed 6/4/2015)

[25] More details are available at: http://www.ntia.doc.gov/category/institute-telecommunication-sciences. (Last accessed November 2015)

- **State and local DOTs, transit agencies, and other infrastructure or fleet operators** will make decisions about investing in, deploying, and maintaining DSRC infrastructure and applications.

- **Academic partners** collaborated with USDOT to develop DSRC and will continue to collaborate to deploy DSRC.

- **Automotive manufacturers** who collaborated on development of the technologies and applications will also include DSRC capabilities that are mandated, and may equip vehicles with optional DSRC technology.

- The **aftermarket** sector may offer DSRC-compatible devices, including mobile devices and applications.

2.IV. Significance of the 5.9 GHz DSRC Band

2.IV.A. DSRC Characteristics Uniquely Suited to Connected Environments

Latency is a crucial factor to crash avoidance. USDOT defines latency as, the length of time between initiation and completion of a process. For the purposes of this discussion, latency is: (a) the time from when an application has a message to send and communicates this to the communications system, to (b) the time the recipient on the other end of the communication system receives the message and provides it to an application for use. As discussed in Section 2.1, DSRC has significant latency advantages over other communications technologies principally because of its one-to-many broadcast mode, which enables all equipped vehicles within range to remain in continuous contact with each other without the need to initiate individual (one-to-one) connections the way other wireless data-exchange protocols do.

Other contributing factors to the latency include propagation delay, message formation, data protocols, routing and switching, and queuing and buffering.

- *Propagation delay*[26] is the time it takes for data to travel from source to destination. This delay happens without regard to transmission size, rate, or protocols involved. It is a function of the transmission distance divided by the transmission speed. The transmission medium (i.e., copper wire, fiber optic cables, coaxial cable, air (radio), etc.) will affect the transmission speed, reducing it from a maximum of the speed of light in a vacuum to the speed of light through the medium involved.

---

[26] Defined at: www.o3bnetworks.com/media/40980/white%20paper_latency%20matters.pdf (accessed May 11, 2015).

- *Message formation delay* is the time it takes to convert bytes of data in storage to packets of data for serial streaming. It is determined by the packet size in bits divided by the transmission rate in bits per second of the communications link used.

- *Data communication protocols* at various levels in the protocol stack use "handshakes" to ensure data properly transfers from one level to another. These "handshakes" focus on acknowledgment of receipt and error detection with errors being noted and information passed back to the sending layer. DSRC eliminates these "handshakes" and therefore eliminates this source of latency.

- *Routing and switching latencies* can accumulate as packets are passed through routers and switches from source to destination. Any outage or congestion on a link along the intended path can alter the path and affect latency. DSRC enables V2V communication without the use of routes and switches, thereby eliminating this potential source of latency.

- *Queuing and buffering latency* refers to the time a packet waits for transmission, dependent on the amount of traffic on the link.

According to the NHTSA Technology Readiness report,[27] DSRC is a technology that provides local, low latency network connectivity. It allows nearly instantaneous network connections as well as broadcast messaging that require no network connection. In addition, DSRC can be considered to be low latency because it consists of point-to-point communication over very short distances (less than 300 meters) with relatively few messaging protocol requirements.

Appendix B presents a more in-depth discussion of latency and offers a comparison of a range of today's communications technologies against the key requirements for crash-avoidance applications. The conclusion to this analysis is that only DSRC meets all of the requirements for V2V safety applications.

---

[27] NHTSA: *Vehicle-to-Vehicle Communications: Readiness of V2V Technology for Application*, DOT HS 812 014, pg. 242, at: http://www.nhtsa.gov/staticfiles/rulemaking/pdf/V2V/Readiness-of-V2V-Technology-for-Application-812014.pdf (Last accessed June 2015)

Low Latency and High Availability

Time is critical in crash avoidance. For example, at 70 miles per hour, a vehicle travels more than 100 feet every second. Thus, some safety-critical V2V applications require that a safety message be delivered at least every 100 milliseconds.[28]

Following the allocation of the 5.9Ghz DSRC band, USDOT worked with industry to develop DSRC technologies that meet rigorous requirements and enable applications that prevent or lessen the impact of a crash with vehicles traveling at high speeds. As a communications medium, DSRC meets the strenuous crash avoidance safety requirements with the following performance characteristics:

- **Speed of transmission:**
  - Very fast network access times.
  - Low latency (delay of well under 100 microseconds).
  - Rapid message delivery (data exchange of over 6 megabytes per second).

- **Limited range:**
  - Broadcast or point-to-point communications over short distances with relatively few messaging protocol requirements (in air, radio transmits information at approximately light speed).[29]
  - Limited range (less than 1,000 meters) to allow spectrum reuse and limit interference.

- **High reliability and stability:**
  - Performance immune to extreme weather conditions (e.g., rain, fog, snow, etc.).
  - Designed to be tolerant to multi-path transmissions typical with roadway environments.
  - Works in high vehicle speed mobility conditions.
  - Offers a stable platform as it does not evolve as rapidly as, for example, cellular options.

---

[28] *System Requirement Description for the 5.9 GHz DSRC Vehicle Awareness Device Specification*, p.72; *The Model Deployment Safety Device DSRC BSM Communication Minimum Performance Requirements-VSC3 Internal Document*, Revision 9.0, 10/10/2011Revision 11.1 01/24/2012; and described in SAE J2735 2009-11/802.11p.

[29] NHTSA: *Vehicle-to-Vehicle Communications: Readiness of V2V Technology for Application*, DOT HS 812 014, at: http://www.nhtsa.gov/staticfiles/rulemaking/pdf/V2V/Readiness-of-V2V-Technology-for-Application-812014.pdf (Last accessed June 2015)

- **Dedication and availability:**

  o DSRC equipment operates in a licensed frequency band that has a primary allocation for transportation safety applications by the FCC. In its 2004 Report and Order, the FCC concludes:

  "...*it is paramount that such communications be protected from interference given the consequences to the traveling public should any one of the safety applications fail due to unacceptable error rates or delay. In this connection, we also agree...that non-public safety use of the 5.9 GHz band would be inappropriate if such use would degrade the safety/public safety applications.*"[30]

Security, Privacy, and No Recurring Subscription Fees

USDOT presumes that participants in a mandatory V2V safety system would not be required to enter into contracts with security or communications service providers or other participants.[31] Because the 5.9 GHz DSRC bandwidth is dedicated to transportation safety, no subscription fees to access the bandwidth would be needed for participation.

In order to protect consumer privacy, USDOT has ensured that no element of the BSM will personally identify the vehicle (as through a Vehicle Identification Number (VIN) or registration number) or the vehicle's driver or owner. The system will not enable tracking through space and time of vehicles linked to specific owners or drivers nor will the safety messages be used by law enforcement or private entities to determine the identity of a speeding or erratic driver. Finally, the system will not collect financial information, personal communications, or other information linked to individuals nor will it create a "pipe" into the vehicle for extracting data. The security solution developed also creates an environment that provides extremely difficult conditions for third parties attempting to use the system to track a vehicle.

USDOT takes very seriously its obligation to protect consumer privacy appropriately, as it supports deployment of V2V DSRC communications. The Department will be publishing a Privacy Impact Assessment (PIA) concurrent with its V2V NPRM that will explain in transparent and clear terms to the public the potential privacy risks created by DSRC communications and the BSM exchanged by motor vehicles, including those stemming from collection of BSMs by third parties not affiliated with the SCMS or "opt-in" service providers authorized by a driver.

---

[30] FCC Report and Order 03-324, February 10, 2004, p.12 at: https://apps.fcc.gov/edocs_public/attachmatch/FCC-03-324A1.pdf (Last accessed June 2015)

[31] http://www.its.dot.gov/connected_vehicle/principles_connectedvehicle_environment.htm. (Last accessed June 2015)

The PIA will detail the technical, physical and policy controls that the Department has designed into the V2V DSRC system that mitigate this and other potential privacy risks related to V2V DSRC communications. For example, in order to minimize potential privacy risks related to collection by third parties of the V2V DSRC communications, the BSM contains no personally identifying information or information that could link a message directly to a driver, motor vehicle owner or the motor vehicle itself (such as VIN or license plate).

BSMs are broadcast in only a limited geographical range so that V2V applications in nearby vehicles can use the information to warn drivers of crash imminent situations. The V2V system, itself, does not collect or store BSMs broadcast by motor vehicles except in the case of apparent equipment malfunction. It would be costly and difficult for a third party monitoring or collecting BSMs to use them to track an individual car or driver. Moreover, other, less expensive methods to do so exist that make it unlikely that a third party would use BSMs for that purpose. NHTSA's V2V PIA will warn consumers that they need to carefully review such Terms of Service prior to providing consent to data collection and explain how to contact that the Federal Trade Commission (FTC) in the event that a service provider violates the Terms of Service. NHTSA also envisions taking steps to ensure that motor vehicle manufacturers provide privacy risk and mitigation information directly to consumers.

In contrast to DSRC, existing commercial communications systems operations (e.g., smartphone data packages or in-vehicle telematics services) are based on pre-existing agreements with their users. User identity resides within databases, which is essential for the user to be able to receive data and services requested when using applications. These identities are also the basis for performing authentication of a user,[32] which supports trusted messaging within these typically closed systems in which all partners are known cohorts. Further, all other wireless communications that could be used for connectivity with vehicles include the use of permanent identifiers (for instance, a Media Access Control (MAC)[33] address or cell phone number) in order to direct the data/services to the user requesting them.

DSRC's ability to provide safety and security with an absence of subscription information databases supports greater levels of privacy, a critical element of public acceptance of Connected Vehicle safety applications. Additionally, USDOT, the Crash Avoidance Metrics Partnership (CAMP), and security experts have designed a leading-edge security system that is uniquely tailored for a highly mobile environment of over 350 million users that does not collect or store personal information.

---

[32] NHTSA: *Vehicle-to-Vehicle Communications: Readiness of V2V Technology for Application*, DOT HS 812 014, at: http://www.nhtsa.gov/staticfiles/rulemaking/pdf/V2V/Readiness-of-V2V-Technology-for-Application-812014.pdf (Last accessed June 2015)
[33] A MAC address is a unique identification associated with a networked device.

DSRC provides a broadcast mode that performs well in crash avoidance situations, whereas other Internet Protocol (IP) media do not. This mode is particularly critical because of the time-critical and location-based[34] needs of the V2V applications. While several technologies have either developed or proposed a "direct" device-to-device mode of communications, they were not designed for constantly changing ad-hoc networks consisting of numerous users traveling at highway speeds. In many cases, these other technologies are not scalable to accommodate Connected Vehicle applications.

USDOT has regularly performed a comparative analysis of communications options:

- From 2002-2012 with CAMP and the Vehicle Infrastructure Integration Consortium (VIIC), research partners representing the Original Equipment Manufacturers (OEMs).[35]
- In 2009 while developing a Report to Congress on the ITS Research Results.
- In 2012 while developing the 2010-2014 ITS Strategic Plan.[36]
- From 2012-2013 in an independent analysis by Booz Allen Hamilton.[37]
- From 2013-2014 in an independent analysis by the American Association of State Highway and Transportation Officials (AASHTO) comparing various media for V2I, I2V, and backhaul capabilities.[38]
- In 2013-2014 in concert with the establishment of the Southeast Michigan 2014 project tests to understand how to transmit data seamlessly across many forms of communications media.[39]
- In a 2014 technical memo to update some of the previous CAMP work to reflect how communications technologies capabilities are evolving.

---

[34] Highly mobile but short-range location-based needs.

[35] The Crash Avoidance Metrics Partnership (CAMP) was formed by Ford and General Motors in 1995 to accelerate the implementation of crash avoidance countermeasures in passenger vehicles to improve traffic safety. The Vehicle Safety Communications Consortium brings together various groups of OEM, including Ford, General Motors, Honda, Mercedes-Benz, Nissan, Toyota and Volkswagen in four cooperative research projects with NHTSA to advance the safety research objectives of the Department's Connected Vehicle research. These research results are summarized in Appendix D of this report and available at:
http://www.nhtsa.gov/DOT/NHTSA/NVS/Crash%20Avoidance/Technical%20Publications/2011/811492A.pdf, p.59. (Last accessed June 2015)

[36] http://www.its.dot.gov/strategic_plan2010_2014/ (Last accessed June 2015)

[37] These research results are summarized in Appendix B of this report and available at:
http://www.its.dot.gov/connected_vehicle/pdf/CDDS_ReadAhead40912v6Final.pdf.

[38] National Connected Vehicle Field Infrastructure Footprint Analysis - Final Report, Appendix A (FHWA-JPO-14-125), AASHTO/USDOT; June 27, 2014 at
http://stsmo.transportation.org/Documents/AASHTO%20Final%20Report%20_v1.1.pdf (Last accessed June 2015)

[39] http://www.its.dot.gov/presentations/pdf/Breakout_7-I_DSRC.pdf, slide 3. (Last accessed June 2015) Additional details from conversations in 2014 with Walt Fehr, ITS Program Manager for Systems Engineering.

Appendix B includes a summary of the recent analyses. With each effort, the conclusion has remained—that DSRC is, to date, the only viable option for safety-critical applications that are location specific and require low latency in the communication channel.

2.IV.B.   Ensuring Competition Among Communication Technologies: Use of non-DSRC Technologies in Connected Vehicle Environments

USDOT reaffirms its intention to explore all wireless technologies for applicability to safety, mobility, and environmental applications. In 2008, the ITS Program framed the definition of connectivity to include both DSRC and non-DSRC technologies as a means of providing a platform for connected transportation environments. Such a framework has multiple benefits:

- It allows for the integration of a wider array of technologies, which enables private industry to develop innovative technologies that offer new or additional features.

- It allows wireless traffic safety system architectures to adapt as technologies evolve over time, ensuring that the networks are able to incorporate improved applications into their existing systems as they become available.

- It ensures that benefits are not limited only to drivers of vehicles equipped with DSRC technologies.

Thus, in addition to DSRC, Connected Vehicle environments will use several types of wireless connectivity—e.g., cellular, satellite, radio, fiber, and Wi-Fi, among others—to serve the public good. Connected Vehicle environments can achieve a more seamless exchange of data from a variety of sources based on new advancements in message protocols that are usable across a range of wireless media. Additionally, new communications technologies are in development, such as 5G or LTE Direct that carry an expectation of near-term availability. USDOT and the ITS Standards community are engaged with these developments to monitor how they can be employed effectively in Connected Vehicle environments:

- Mobility management will be improved when drivers, transit riders, and freight managers all have access to substantially more real-time, accurate, and comprehensive information on travel conditions and options; and when system operators, including roadway agencies, public transportation providers, and port and terminal operators, have actionable information and the tools to affect the performance of the transportation system in real-time. Actionable information may also increase safety; e.g., with road segment specific weather conditions.

- Environmental impacts of vehicles and travel can be reduced when travelers can make informed decisions about modes and routes and when vehicles can communicate with the infrastructure to enhance fuel efficiency by avoiding unnecessary delays and stops.

Based on USDOT and partner research, the existing range of communications technologies can effectively support applications that do not require low latency, high reliability and availability, or privacy (since many of them are based on users to opting-in to services). Results are summarized in Appendix B and documented in:

- A Footprint Analysis performed by AASHTO that investigated how Connected Vehicle implementations could leverage existing communications technologies.[40]

- Concepts of Operations documents for the range of applications described above[41].

- Ongoing prototype testing such as:

  o The Southeast Michigan 2014 Project, which offers a "plug and play" opportunity for vendors and application developers to test their products. If the products are developed correctly based on the Connected Vehicle standards, "plugging" into the Southeast Michigan systems will prove an important level of interoperability. Notably, the Southeast Michigan test bed has been designed using multiple communications media and is advancing new data protocols that will allow data to be exchanged seamlessly over any communications media.

  o Prototyped Dynamic Mobility Applications that are being built and tested in an iterative process at operational sites around the nation:

    ▪ Three sites are developing and testing different features of the Freight Advanced Traveler Information Systems (FRATIS) applications bundle. [42] The participating vehicles and infrastructure elements will employ Bluetooth and Wi-Fi, and the dispatching systems will continue operating on their existing communications (predominantly cellular). Whatever the medium, data will be exchanged in a secure manner with limited access and use after it is collected.

    ▪ One site in Arizona is developing and testing the Multi-Modal Intelligent Traffic Signal Systems (MMITSS)[43] applications bundle using DSRC for the applications and data exchange within the intersection; and Ethernet and T1 backhaul connections for management from the traffic operations center.

---

[40] *National Connected Vehicle Field Infrastructure Footprint Analysis - Final Report (FHWA-JPO-14-125)*, AASHTO/USDOT; June 27, 2014 at
http://stsmo.transportation.org/Documents/AASHTO%20Final%20Report%20_v1.1.pdf (Last accessed June 2015). Additional materials can be found at: http://stsmo.transportation.org/Documents/Executive%20Briefing.pdf and at http://stsmo.transportation.org/Documents/Task%206a%20AASHTO_CV_Footprint_Deployment_Scenarios_v2.pdf. (Last accessed June 2015).
[41] Available on the www.its.dot.gov website or in the National Transportation Library at: www.ntl.bts.gov.
[42] For more information on FRATIS see: Cambridge Systematics, Inc. *Freight Advanced Traveler Information System (FRATIS) Concept of Operations (ConOps),* Final Report for FHWA, June 13, 2012.
[43] For more information MMITSS see: University of Arizona (lead); *MMITSS Final Concept of Operations*; Version 3.1; December 4, 2012. http://www.cts.virginia.edu/wp-content/uploads/2014/05/Task2.3._CONOPS_6_Final_Revised.pdf (Last accessed June 2015).

To understand the range of applications that can utilize alternative communications media, the Connected Vehicle Environment applications in which USDOT has invested are listed, below. [44] They include roadway, transit, accessible transportation, and freight management applications; road weather motorist alert and advisory applications; and applications to support management strategies for improved environmental transportation performance. Other private sector applications are under development in anticipation of deployment.

Mobility:

- Advanced Traveler Information Systems

- Intelligent Traffic Signal System (I-SIG)

- Signal Priority (transit, freight)

- Mobile Accessible Pedestrian Signal Systems (PED-SIG)

- Emergency Vehicle Preemption (PREEMPT)

- Dynamic Speed Harmonization (SPD-HARM)

- Queue Warning (Q-WARN)

- Incident Scene Pre-Arrival Staging Guidance for Emergency Responders (RSP-STG)

- Incident Scene Work Zone Alert for Drivers and Workers (INC-ZONE)

- Emergency Communications and Evacuation (EVAC)

- Connection Protection (T-CONNECT)

- Dynamic Transit Operations (T-DISP)

- Dynamic Ridesharing (D-RIDE)

- Freight Specific Dynamic Travel Planning & Performance

- Drayage Optimization

- Accessible Transportation

Environment:

- Eco-Approach and Departure at Signalized Intersections

- Eco-Traffic Signal Timing

---

[44]The full list can be found at: http://www.its.dot.gov/pilots/cv_pilot_plan.htm. (Last accessed June 2015)

- Eco-Traffic Signal Priority
- Connected Eco-Driving
- Wireless Inductive/Resonance Charging
- Eco-Lanes Management
- Eco-Speed Harmonization
- Eco-Cooperative Adaptive Cruise Control
- Eco-Traveler Information
- Eco-Ramp Metering
- Low Emissions Zone Management
- AFV Charging/Fueling Information
- Eco-Smart Parking
- Dynamic Eco-Routing (light vehicle, transit, freight)
- Eco-ICM Decision Support System

Smart Roadside:

- Wireless Inspection
- Smart Truck Parking

Road Weather:

- Motorist Advisories and Warnings (MAW)
- Enhanced Maintenance Decision Support Systems (MDSS)
- Vehicle Data Translator (VDT)
- Weather Response Traffic Information (WxTINFO)

Agency Data:

- Probe-Based Pavement Maintenance
- Probe-Enabled Traffic Monitoring
- Vehicle Classification-based Traffic Studies
- CV-Enabled Turning Movement & Intersection Analysis
- CV-Enabled Origin-Destination Studies
- Work Zone Traveler Information

Some of the Connected Vehicle Environment applications listed above might employ more than one form of communications during use. For instance, traffic signal applications might utilize

DSRC when vehicles or portable devices are within intersection range, especially if the intersection is particularly complex or there is a need to alert drivers to emerging threats. However, for basic knowledge about the signal phase and timing along a corridor, local Wi-Fi might be acceptable. Note that in the former example, the use of DSRC will require security credentials; but when the information is broadcast by other communications modes as an informational broadcast only, it may or may not include security and trust credentials.

Since many public sector agencies (State DOTs, transit agencies, or public safety and emergency management agencies, among others) and private sector organizations (tolling authorities, nationwide freight logistics companies, business or hospital campuses, etc.) have invested in and installed communications technologies to support transportation management and provide services (e.g., logistics, automated vehicle location (AVL), traveler information, or field-to-management center backhaul), FHWA is leading development of Guidance on how to leverage these existing systems and services for Connected Vehicle environments. Appendix C offers descriptions of the many Connected Vehicle test beds and pilot deployment sites that have emerged in the US and around the world in the last 10 years.

To use cellular and other types of (non-DSRC) communication media in Connected Vehicle environments requires additional advancements—USDOT has launched development of new message protocols and security credential management solutions that will allow seamless data exchange across the range of these technologies:

- **Security**—Non-DSRC communications media will need to provide security credentials in the same or a compatible manner with Connected Vehicle security to ensure that Connected Vehicle messages will be authenticated and trusted. In today's communications services, there exists a range of security options from no security to proprietary forms of security. Consumers opt-in to using applications that allow data transfers based on their acceptance of the form and level of security. Connected Vehicle safety applications, in particular, will require a higher level of security (and privacy) and a more rigorous certification process than today's devices and applications used with social media.

- **Reliability**—Service providers making use of communications media other than DCRC may need to demonstrate how they can provide reliable and timely data messaging, especially under congested conditions. For example, mobility applications related to vehicle emissions reductions and fuel consumption reductions will need frequent communications with traffic signals in order to provide drivers actionable feedback on how to time their approach and departure at intersections. The several seconds of delay or momentary unavailability that may be possible with cell phone communication under highly congested conditions would compromise or disable such applications;

- **Privacy**—USDOT may choose to issue Data Privacy Policy best practices applicable to service providers offering optional or "opt-in" applications to ensure that consumer, prior to "opting in," fully understand the application's Terms of Use/Data Privacy Policies: specifically, what data will be collected, how it will be used, with whom it will be shared, how it will be secured, and for how long it will be kept prior to disposal. NHTSA's V2V PIA will likely warn consumers that they need to carefully review such Terms of Service prior to providing consent to data collection by third party service providers and explain how to contact that the FTC in the event that a service provider violates the Terms of Service. NHTSA also envisions taking steps to ensure that motor vehicle manufacturers provide privacy risk and mitigation information directly to consumers.

The same basic safety message (BSM) that has been developed to support V2V safety applications in motor vehicles also will enable a wide range of applications by service providers offering mobility, environmental and additional safety benefits to consumers, some but not all of which will be "opt-in". Many of these applications are categorized as "V2I applications" although they do not always require information obtained from infrastructure. These are in addition to the V2I safety applications, which do require information from the infrastructure, such as the signal phase and timing (SPaT) message.

Typically, the potential risks to individual privacy stemming from exchange of the BSM in V2V technologies also exist in connection with V2I optional and "opt-in" applications. However, unlike V2V applications integrated into a motor vehicle by its manufacturer, optional and "opt-in" V2I applications offered by private or local public entities may involve more risk to individual privacy than do V2V applications in motor vehicles, standing alone. That is because, in order to "opt in" to some applications, consumers will need to provide service providers with personally identifying information not required for (or permitted to be collected in connection with) V2V applications integrated into motor vehicles, and may use additional data elements in the BSM not part of the mandatory BSM broadcast envisioned by NHTSA. The data collected by such authorized service providers often is linked directly to a subscriber (i.e. an individual or his or her vehicle), resulting in a privacy risk profile that is quite different from that of V2V technologies (which are designed specifically to prevent any such linkage). However, it also is important to note that consumers wishing to take advantage of such V2I services will do so voluntarily and perhaps even pay a fee. In so doing, consumers will have the opportunity to review an application's *Terms of Use* and be required to provide consent to the data collection and data privacy terms offered by the service provider.

NHTSA's V2V PIA will likely warn consumers that they need to carefully review such *Terms of Use* prior to providing consent to data collection and explain how to contact the FTC in the event that a service provider violates the *Terms of Use or Terms of Service*. NHTSA also envisions taking steps to ensure that motor vehicle manufacturers provide privacy risk and mitigation information directly to consumers. In addition, FHWA is evaluating these issues and will publish best practices guidance on data access and data privacy in the V2I system.

2.V. Benefits of Connected Vehicle Applications to the Nation

Improved safety of the national transportation system is a critical benefit to the deployment of Connected Vehicle applications.

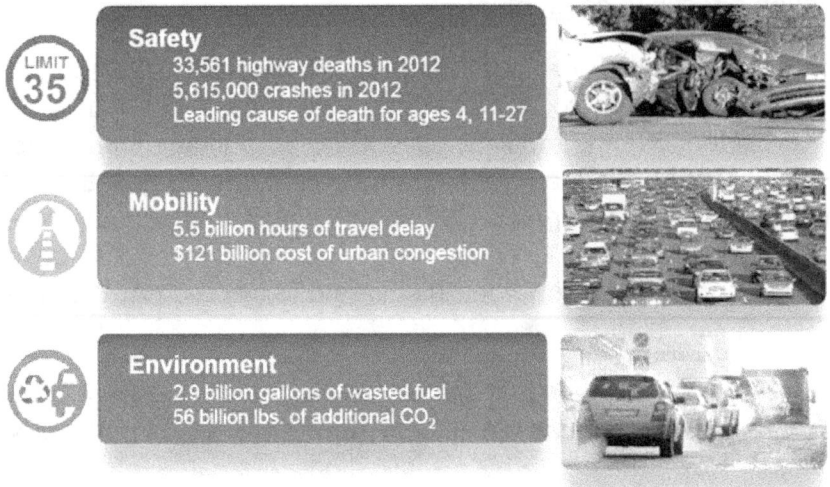

Graphic: USDOT

As noted earlier, NHTSA estimates that if V2V technologies <u>alone</u> are widely deployed, they have the potential to address 81 percent of light-vehicle crashes involving unimpaired drivers by providing warnings to drivers.[45] Overall, USDOT analyses concluded that, as a primary countermeasure, a fully mature V2V system could potentially address the following unimpaired crashes (denoted "target crashes" below):[46]

- About 4,409,000 police-reported or 79 percent of <u>all vehicle</u> target crashes.
- 4,336,000 police-reported or 81 percent of all <u>light-vehicle</u> target crashes.

---

[45] See original analysis at:
http://www.nhtsa.gov/DOT/NHTSA/NVS/Crash%20Avoidance/Technical%20Publications/2010/811381.pdf, pg. vi
[46] Results summarized from two reports: *Frequency of Target Crashes for IntelliDrive Safety Systems* (Najm, Koopman, Smith, and Brewer, October 2010, DOT HS 811 381). See:
http://www.nhtsa.gov/Research/Crash+Avoidance/ci.Office+of+Crash+Avoidance+Research+Technical+Publications.print (last accessed June 2015); and *Analysis of Light Vehicle Crashes and Pre-Crash Scenarios Based on the 2000 General Estimates System* (Najm, Sen, Smith, and Campbell, Nov. 2002, DOT HS 809 573). See:
http://www.nhtsa.gov/Research/Crash+Avoidance/ci.Office+of+Crash+Avoidance+Research+Technical+Publications.print (last accessed June 2015).

- 267,000 police-reported or 81 percent of all <u>heavy-truck</u> target crashes annually.[47]

The analysis also indicated V2I systems could potentially address the following unimpaired crashes[48]:

- About 1,465,000 police-reported or 26 percent of <u>all-vehicle</u> target crashes.
- 1,431,000 police-reported or 27 percent of all <u>light-vehicle</u> target crashes.
- 55,000 police-reported or 15 percent of all <u>heavy-truck</u> target crashes annually.

In addition, combined V2V and V2I systems could potentially address the following unimpaired crashes[49]:

- About 4,503,000 police-reported or 81 percent of <u>all-vehicle</u> target crashes.
- 4,417,000 police-reported or 83 percent of all <u>light-vehicle</u> target crashes.
- 272,000 police-reported or 77 percent of all <u>heavy-truck</u> target crashes annually.[50]

Recent NHTSA analysis finds that there are three safety applications that the agency believes are enabled by DSRC-based vehicle-to-vehicle communications and could not be replicated by any current, known vehicle-resident sensor- or camera-based systems, as discussed below.

1. **Intersection Movement Assist (IMA):** Intersection Movement Assist warns the driver of a vehicle when it is not safe to enter an intersection due to a high probability of colliding with one or more vehicles at intersections both where a signal is present (a "controlled" intersection) and those where only a stop or yield sign is present (an "uncontrolled" intersection).

2. **Left Turn Assist (LTA):** Left Turn Assist warns the drivers of vehicles, when they are entering an intersection, not to turn left in front of another vehicle traveling in the opposite direction.

3. **Emergency Electronic Brake Light (EEBL):** Emergency Electronic Brake Light enables a vehicle to warn its driver to brake in a situation where another V2V-equipped vehicle decelerates quickly but may not be directly in front of the warning vehicle. The EEBL warning is particularly useful when the driver's line of sight is obstructed by other vehicles or bad weather conditions, such as fog or heavy rain.

---

[47] NHTSA: *Vehicle-to-Vehicle Communications: Readiness of V2V Technology for Application*, DOT HS 812 014, page 18-19, at: http://www.nhtsa.gov/staticfiles/rulemaking/pdf/V2V/Readiness-of-V2V-Technology-for-Application-812014.pdf (Last accessed June 2015)

[48] Ibid.

[49] Ibid.

[50] Ibid.

NHTSA provides a preliminary estimate of the non-discounted annual maximum benefits, assuming that all passenger vehicles (PVs) are equipped with only IMA and LTA and the communication rate reaches 100 percent among PVs as follows:[51]

- 412,512 to 592,230 crashes prevented;
- 777 to 1,083 lives saved;
- 191,202 to 270,011 MAIS 1-5 injuries reduced; and[52]
- 511,118 to 728,173 property-damage-only vehicle incidents (PDOVs) prevented.

In addition, a recent report[53] projects the potential safety benefits of crash warning applications using V2V communications on board light vehicles (e.g., passenger cars, vans and minivans, sport utility vehicles, and light pickup trucks with gross vehicle weight rating under 10,000 pounds). The four selected V2V crash-warning applications (Forward Collision Warning, Intersection Movement Assist, Left Turn Assist, and Blind Spot Warning/Lane Change Warning) have an estimated 49 percent crash avoidance effectiveness, reducing about 1,453,000 police-reported crashes annually. The reduction in crashes is approximately 30 percent of all annual police-reported crashes within the United States that involved at least one light vehicle. This analysis focuses on light vehicles as the host vehicles of the safety applications, and assumes the following:

- All light vehicles are fully integrated with V2V technology and corresponding safety applications.
- All vehicles (i.e., light vehicles, medium and heavy trucks, busses, and motorcycles) are fully equipped with V2V technology and can communicate with fully integrated light vehicles.
- V2V safety applications are always available to perform their safety functions.
- V2V safety applications have 100 percent accuracy; i.e., identification of a true driving conflict and accurate warning activation.
- Simple driving conflicts are modeled using basic kinematic equations, where the host vehicle only responds to the conflict while the other vehicle stays the course.

---

[51] Of the above estimated benefits, IMA would prevent 310,451 to 416,458 crashes, save 671 to 900 lives, reduce 136,959 to 176,593 Maximum Abbreviated Injury Scale (MAIS) 1-5 injuries, and eliminate 399,431 to 535,823 PDOVs. LTA would avoid 102,061 to 175,772 crashes, save 106 to 183 lives, reduce 54,243 to 93,418 MAIS 1-5 injuries, and eliminate 111,687 to 192,350 PDOVs. Located in NHTSA research publication *Vehicle-to-Vehicle Communications: Readiness of V2V Technology for Application* p. 287, at: http://www.nhtsa.gov/staticfiles/rulemaking/pdf/V2V/Readiness-of-V2V-Technology-for-Application-812014.pdf

[52] MAIS represents the maximum injury severity of an occupant at an Abbreviated Injury Scale (AIS) level. AIS ranks individual injuries by body region on a scale of 1 to 6: 1=minor, 2=moderate, 3=serious, 4=severe, 5=critical, and 6=maximum (untreatable).

[53] NHTSA research report *Estimation of Safety Benefits for Light-Vehicle Crash Warning Applications Based on Vehicle-to-Vehicle Communications* will be published in late 2015.

- No external conflicts or unintended consequences are modeled, such control loss or secondary impacts.

- Crash impacts are inelastic; i.e., center-of-mass collisions, including conservation of momentum, without incorporating any additional resulting impacts.

- The "host" vehicle is the light vehicle, the host vehicle initiates the applicable maneuver.

- Driver of the host vehicle responds to a driving conflict (or warning) with a single appropriate response (brake or steer) that depends on the nature of the conflict. The host driver is not impaired.

- All motion and reaction occur without intermittent delays or interference, and are constant until otherwise acted upon.

These general assumptions are carried throughout the application of the safety benefits estimation methodology, independent of the V2V safety application or pre-crash scenario.

Similarly, FHWA analysis notes that red light-running crashes may truly be addressable only through Connected Vehicle environments because of reliance upon fast and secure transmission of the signal phase and timing data in the event of an emerging violation. With assurances that this data is broadcast in a trusted and reliable manner, the violating vehicle can authenticate the source and use the information to calculate whether the driver will violate the red light. In addition, the local traffic signal system can use the same information to hold the red light for all approaches to ensure that no one else enters the intersection.

A 2010 study estimated that 229,333 red light-running crashes occur annually, and that Connected Vehicle applications offer the capability to prevent these types of crashes.[54] An update to this study was completed in draft form in December 2014, and will be incorporated into a larger study on V2I safety applications for publication in late 2016, or early 2017. It evaluates the number of red-light violation crashes that may be prevented each year using the NHTSA model of DSRC penetration into the light vehicle fleet with the AASHTO estimate of DSRC-equipped traffic signals.

---

[54] *Crash Data Analyses for IntelliDrive Vehicle-Infrastructure Communications for Safety Applications,* Revised Final Report prepared by Vanasse Hangen Brustlin, Inc. for the Federal Highway Administration, p. 46.

Chapter 3 Status of DSRC through Research and Development

To provide the context for assessing the status of DSRC technologies, applications, and use of the DSRC medium, this chapter presents:

3.I. An overview of the enabling DSRC technologies (both in-vehicle and infrastructure-based):
  A. Essential Vehicle-Based Components
  B. Essential Infrastructure Components

3.II. An overview of the critical V2V and V2I crash avoidance applications that require DSRC for operations inclusive of:

3.III. An introduction to DSRC and V2V Communications Security including an overview of:
  A. The Security Credential Management System
  B. System Integrity and Management
  C. Organization and Ownership
  D. System Governance

3.IV. An identification of the essential standards that: support successful V2V and V2I DSRC operations and spectrum utilization; enable interoperability among varying makes and models of devices; ensure message interoperability; and provide secure, authenticated, and trusted messaging.

3.V. Conclusions regarding the maturity and status of DSRC:
  A. Assessment of Status and Maturity of the DSRC Connected Vehicle Technologies and Applications
  B. Assessment of Status and Maturity of the DSRC Implementation
  C. Analysis of known and potential gaps
    i. Coexistence with Unlicensed Users
    ii. Frequency Coordination
    iii. Open Items
  D. An assessment by the GAO on V2V and DSRC readiness
  E. Relationship of DSRC to emerging Automated Vehicle technologies

3.I. Overview of DSRC Enabling Technologies

USDOT and automobile industry efforts to develop Connected Vehicle technologies have focused on both in-vehicle components and external, infrastructure components. These components include a security system that manages the credentials that result in trusted and authenticated data being broadcast and exchanged among vehicles and between vehicles and roadside infrastructure. A number of standards have also been developed for how these components function and interact. This section summarizes the internal and external components as well as the Cooperative System Standards.

3.I.A. Essential Vehicle-Based Components

Essential in-vehicle components include:

- **Hardware:** DSRC radios, cables, and antennae to gain wireless connectivity and support data exchange with other vehicles and V2I safety roadside infrastructure, and GPS chips used to determine vehicle location and time. Other hardware includes the secure memory and microprocessor to store security credentials and perform the applications processing.

- **Software:** Software applications that receive, process, and analyze timestamped data broadcast by other vehicles, such as GPS location, speed, heading and brake status and, based on that analysis, predict when collisions are imminent. Software also includes applications that analyze and process data communicated from V2I safety roadside infrastructure.

- **Driver-Vehicle Interface (DVI):** A DVI that—based on the data analysis conducted by the V2V and V2I software applications—provides a warning to the driver when a collision may be imminent, through vehicle-based features such as sounds, lights, or seat vibrations.[55] The alerts and warnings provided by the DVI will be compliant with the NHTSA guidelines on Driver Distraction.[56]

- **Data/Basic Safety Message:** The BSM is a key data element in vehicle-to-vehicle communications. The BSM communicates a vehicle's essential state information to a set of neighboring vehicles over a given range. It is broadcast from the vehicle every 100 milliseconds. Vehicles in range that are equipped with safety applications process the incoming BSM data to determine the threat of collision with the sending vehicle under a variety of scenarios (e.g., blind spot, intersection, forward path collision). The BSM includes the large set of common data elements needed by various safety applications while avoiding the need for application-specific messages and saving over-the-air bandwidth.[57]

Figure 3-1 on the following page illustrates these essential technologies on the vehicle.

---

[55] For example, for potential collisions involving vehicles in a driver's blind spot, one automobile manufacturer's vehicles provided three short low-pitched beeps repeated three times, an orange light in the side view mirror, and a vibration on the side of the driver's seat in the direction of the potential collision.

[56] Guidelines are located at: http://www.nhtsa.gov/Research/Human+Factors/Distraction (Last accessed June 2015)

[57] Definition found in *Vehicle Safety Communications – Applications VSC-A Second Annual Report,* January 1, 2008 through December 31, 2008, p. 17, located at:
ttp://www.nhtsa.gov/DOT/NHTSA/NVS/Crash%20Avoidance/Technical%20Publications/2011/811466.pdf. (Last accessed June 2015)

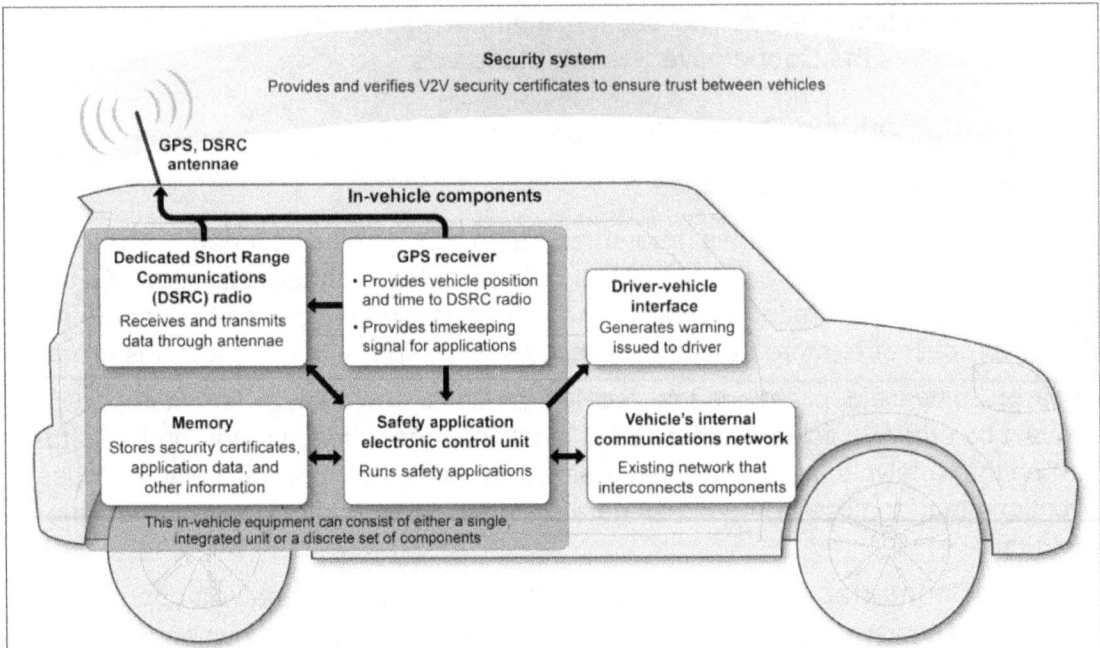

Sources: Crash Avoidance Metrics Partnership and GAO.

Figure 3-1. Illustration of DSRC Vehicle-Based Technologies

A range of DSRC vehicle-based devices can be produced with varying levels of functionality, thus supporting flexibility in deployment planning. The range includes:

- **Integrated on-board equipment (OBEs)** are devices that offer the most sophisticated capabilities due to their direct integration into the vehicle and their access to data drawn from the vehicle's controller area network bus and other systems to support crash avoidance safety applications.

- **Retrofit equipment** are devices added after the vehicle manufacturing process and are formally installed by authorized technicians to ensure that the integration with the vehicle systems does not cause harm or malfunction. Although integration can occur at many levels, retrofit devices have access to data and vehicle-based driver interfaces that aftermarket devices typically do not include. These devices perform in a manner most similar to OBEs.

- **Aftermarket Safety Devices (ASDs)** are devices added to a motor vehicle after assembly. The level of connection to the vehicle can vary based on the type of aftermarket device itself. For example, a "self-contained" V2V ASD could only connect to a power source, and otherwise would operate independently from the systems in the vehicle. The self-contained device would include a DVI for advisories and warnings. Aftermarket V2V devices can be portable devices inserted into a fixed cradle in the vehicle (e.g., cell phones with applications), or added to a vehicle at a vehicle dealership or by authorized dealers or installers of automotive equipment.

- **Vehicle Awareness Devices (VADs)** are the simplest design. A VAD only transmits a BSM to nearby vehicles. A VAD does not have any safety applications or driver interface; and it cannot provide any advisories or warnings to a driver. Installing these devices on existing vehicles could be an attractive option for fleet operators, rental agencies, or vehicle owners who could see benefit in signaling the presence of their vehicles to V2V-equipped vehicles and thus potentially avoiding crashes. Installation of VADs could increase deployment of V2V systems across the fleet as a whole, and thus potentially could increase the benefits for early adopters of this technology.

Detailed definitions for these devices that illustrate the differences in their capabilities can be found in Appendix E.

During the Safety Pilot Model Deployment project over 3,000 participants were equipped with a range of these devices—sixty-four vehicles were equipped with integrated OEM solutions (a CAMP-developed device), 300 vehicles had aftermarket technology installed, 19 heavy vehicles (16 trucks and 3 transit buses) were retrofitted with equipment, and 2,850 vehicles were outfitted with vehicle-awareness devices, which can transmit the BSM to other vehicles but cannot receive information needed to alert the driver. Many of the systems had internal components designed and built by a number of different manufacturers and suppliers, illustrating interoperability across makes and models. With these different devices operating together as a system, providing alerts and advisories to drivers, USDOT has generated a representation of how a fully functional V2V system might work. Additionally, the results on interoperability provide USDOT, OEMs, and State and local agencies with a basis for planning for "day-one" benefits knowing that a range of devices can be deployed together to support the process of transforming safety, mobility, and environmental performance.

3.I.B.   Essential Infrastructure Components

To enable communication between vehicles and roadside equipment and support V2I applications, DSRC must be integrated with existing traffic equipment (e.g., traffic signal controllers or backhaul connections to Traffic Management Centers (TMCs)). All of these components are considered "roadside equipment" (RSE) or "roadside infrastructure". These terms describe categories of infrastructure components that include DSRC-based roadside units (RSUs), which are installed as part of the RSE. DSRC RSUs are essentially devices with radios and act as data processors that facilitate communication between vehicles and other devices with the roadside infrastructure by exchanging data over the 5.9 GHz DSRC band. As illustrated in Figure 3-2, examples of roadside infrastructure components include:[58]

- **Infrastructure communications equipment**: Including DSRC radios that allow the V2I safety infrastructure applications platform and other Connected Vehicle services (e.g., security services) to communicate wirelessly at the necessary latency and network attach time to exchange information with other DSRC-equipped devices. Roadside

---

[58] Based on: http://ntl.bts.gov/lib/48000/48500/48527/ED89E720.pdf, page 79. (Last accessed June 2015)

communications equipment also includes other elements not requiring low latency wireless transmission that link the infrastructure applications platform with local supporting equipment such as traffic signal controller units, and link to off-site, back office elements such as TMCs or security management organizations. This is known as backhaul communications.

- **Applications:** An infrastructure application platform that hosts V2I safety applications, receives information relevant to those applications from roadside detectors and traffic signal controller units, and communicates with the infrastructure communications equipment.

- **Data Sensors and Other Equipment:** Infrastructure data equipment, such as traffic signal controller units, weather detectors, pedestrian detectors, and traffic detectors that provide information to the V2I safety application platform.

- **Driver-Vehicle Interface (DVI):** A DVI that uses the data analysis conducted by the V2V and V2I software applications to provide a warning to the driver through infrastructure-based devices such as a changeable, dynamic message sign that is compliant with the Manual on Uniform Traffic Control Devices (MUTCD) to alert drivers to upcoming hazards or dangerous conditons.

Figure 3-2 on the next page illustrates one example of the interaction between vehicle- and infrastructure-based components. There will be other ways to perform these functions, and additional functions that may be added, but this illustrates a typical configuration at many V2I deployments.

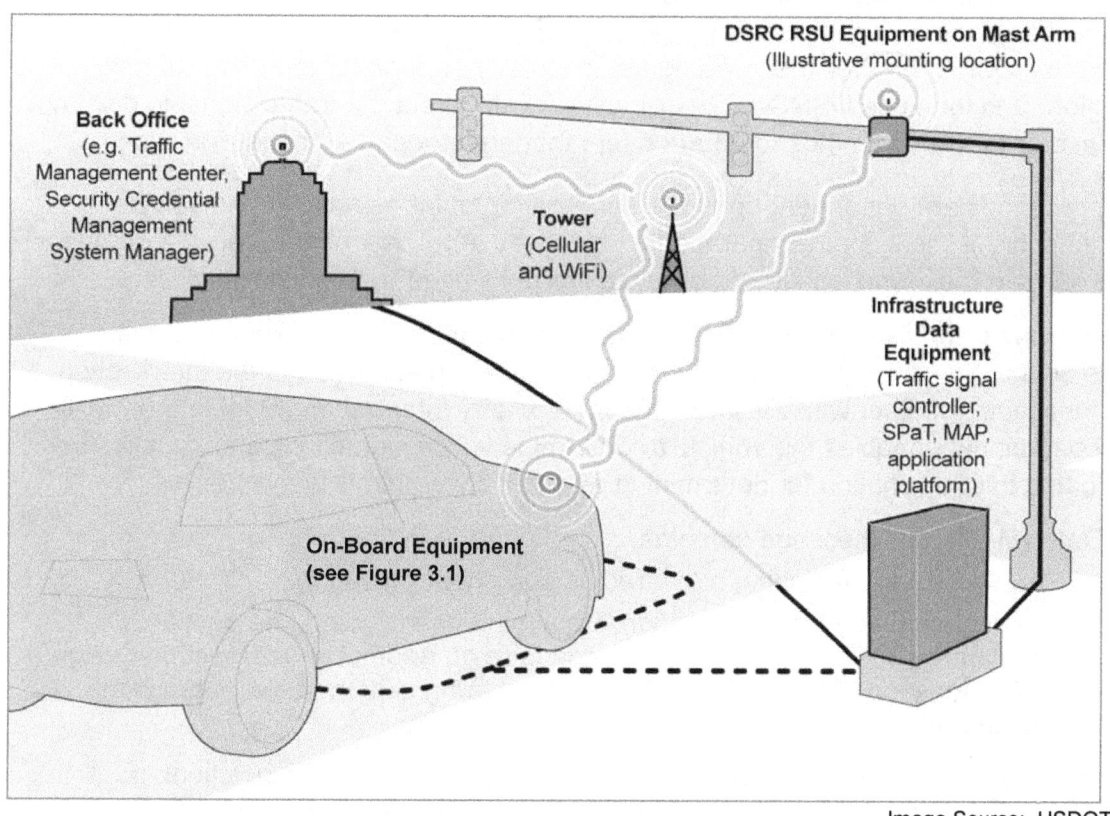

Image Source: USDOT

Figure 3-2. Basic External V2V and V2I Roadside Infrastructure Components for a Typical Deployment

3.II. Overview of the DSRC Crash Avoidance Safety Applications

The DSRC technologies are specifically designed to support V2V and V2I crash avoidance safety applications requirements (as described previously in Table 2-1 which lists the current V2V and V2I safety applications that require DSRC. See Appendix F for more details). Given the innovative nature of the application marketplace, USDOT believes that additional safety applications will be developed in the future.

The same wireless technology that supports V2V safety applications (5.9 GHz DSRC) also enables a broader set of safety applications when combined with compatible roadway infrastructure. V2I communications involve the wireless exchange of critical safety and operational data between vehicles (including personal electronic devices) and highway infrastructure, intended primarily to avoid motor vehicle crashes while enabling a wide range of mobility and environmental benefits.

Similar to V2V, some V2I messages are expected to require DSRC to broadcast from the infrastructure because they require low latency and rapid network association time in a highly mobile environment (see the list below). However, some V2I safety applications may employ DSRC in a manner that evolves throughout operation. For instance, a red-light violation system warning could use DSRC to notify the violating vehicle (early enough for the driver to take corrective action) and then use DSRC, LTE, or local Wi-Fi to direct the traffic signal to display an all-red phase or to relay warnings to other drivers that are stopped at the intersection.

V2I safety applications are dependent upon the cooperation of infrastructure and vehicle components to achieve the system's operational objectives. Applications that require broadcast messages to support their functionality include, but are not limited to, the following:

- **Signal Phase and Timing (SPaT):** this message contains information and current status on the phase and timing of all the signals for each approach in the intersection. This message, together with the intersection geometry information (MapData or "MAP", described below), enables the vehicle to determine which signal indication applies to it and to use this information for determining whether a warning is warranted.

- **MapData (MAP):** this message contains:
  - Road geometry, including curve radius and super-elevation.
  - Intersection Geometry Information: Detailed data including intersection ID, road/lane geometry for all approach roads (e.g., geometric intersection design or "GID"), location of stop lines, and lane numbering scheme associated with movements.

- **Positioning correction** (optional): global positioning system (GPS) positioning correction information for the intersection that the vehicle may use to improve its estimate of location within the intersection.

- **Road surface information and other weather-related data if available** (optional): information about the road surface coefficient of friction at the intersection; weather related data such as dew point, temperature, visibility, and rain, which may assist the vehicle components in adjusting warning timing to account for variations in stopping distance.

Implementation of these V2I applications will require additional data elements to be broadcast to, and processed by, vehicles. Since the broadcasting of additional data has the potential of leading to communication congestion, USDOT's ITS JPO is conducting additional channel congestion analysis to ensure that V2V and V2I safety messaging is not compromised due to broadcasting more data for V2I non-safety applications.[59] As described in the next section, BSM data, and potentially the SPaT and MAP data, may utilize the safety channel—Channel 172. Service channels 174 and 178 are the channels available for non-safety data exchange

---

[59] NHTSA: *Vehicle-to-Vehicle Communications: Readiness of V2V Technology for Application*, DOT HS 812 014, page 52, at: http://www.nhtsa.gov/staticfiles/rulemaking/pdf/V2V/Readiness-of-V2V-Technology-for-Application-812014.pdf (Last accessed June 2015)

(see Appendix I for a description of 5.9 GHz DSRC spectrum operations). Use of all of the channels is currently being tested and will be demonstrated in the Southeast Michigan 2014 project. Affiliated test bed sites and future Connected Vehicle pilot sites will also demonstrate use of the band in this manner.

3.III. DSRC and V2V Communications Security[60]

In order for the information in a V2V communication to be useful for safety-critical functions, it must be received in a timely manner, it must be reliable, and it must be transmitted in a standard format. Vehicles participating in the V2V communications network communicate with other connected vehicles using standardized DSRC message types broadcast on a standardized network over a standardized wireless layer. [61,62] V2V communications consist of two message parts: the security credential and the safety message. The safety messages are used to support the safety applications, and the credential ensures that each transmission is from a trusted source.

The BSM broadcast by each DSRC-equipped vehicle includes information about the vehicle's behavior such as its location, its predicted path, its lateral and vertical acceleration, and its yaw rate. Each BSM is time-stamped so the receiving vehicles know when the data elements in the message were created. The BSM contains no personally identifying information (PII) and is broadcast in a very limited geographical range, typically less than 0.5 km. Motor vehicles nearby will receive and analyze the information in order to warn drivers of crash-imminent situations. Except in the case of "misbehavior," the system will not collect, and motor vehicles will not store, complete sets of messages sent or received by V2V devices. The BSM is split into two parts to guarantee that the core information for vehicle safety (Part I) has priority and is transmitted more often. This also reduces the amount of data communicated between devices, helping to reduce channel congestion.

As noted in Chapter IX of NHTSA's Technology Readiness report,[63] the foundation of the V2V system is "trust". In this case, trust is defined by the requirement that thousands of data messages will be authenticated, in real-time, as coming from a trusted (but unknown) source. It is also a critical element in achieving interoperability—the ability of vehicles of different makes, models, and years to exchange trusted data without pre-existing agreements or significant

---

[60] ibid pages 73-73
[61] Per IEEE 1609.4
[62] Per IEEE 802.11
[63] NHTSA: *Vehicle-to-Vehicle Communications: Readiness of V2V Technology for Application*, DOT HS 812 014, at: http://www.nhtsa.gov/staticfiles/rulemaking/pdf/V2V/Readiness-of-V2V-Technology-for-Application-812014.pdf (Last accessed June 2015)

alteration of the existing vehicle designs. Further, the system must be secure against internal and external threats or attacks.

Fundamental requirements for the V2V security system design are that it:

- Appropriately protects privacy (does not require the identity of the participating parties);

- Is fast enough to fit within the bandwidth constraints of DSRC and the processing constraints of the V2V on-board equipment;

- Uses a number of over-the-air bytes for security that fits within the constraints of DSRC bandwidth and the size of the BSM in the message payload; and

- Supports non-repudiation.

NHTSA's current research indicates that the V2V system will use a Public Key Infrastructure (PKI) to authenticate messages, so that vehicles are able to trust the messages they receive. PKI uses digital credentials to inform a receiving device that the message is from a trusted source, and, when appropriate, it uses cryptography to encrypt message content. For V2V communications, BSM messages are trusted but not encrypted, while messages that contain security information (e.g., credentials) are both trusted and the contents encrypted. By using digital credentials to establish trust, rather than encrypting the BSM, sending it, and then requiring the other vehicle to receive, decrypt, and translate it, the system is more efficient and able to achieve the extremely fast communication that a crash-imminent warning application requires.[64]

3.III.A.   Security Credential Management System (SCMS)

The SCMS under development for V2V communications uses asymmetric cryptography where two keys are mathematically linked. The keys function in such a way that what is encrypted with one key can only be decrypted with the other. This property allows one key, the "public key," to be widely distributed while the other key, the "private key," is held only by the owner and cannot be used to determine easily the public key. This is what makes asymmetric cryptography a superior choice over symmetric cryptography (where the same key is used to encrypt and decrypt messages).

The security system design is based on a hierarchical PKI containing a Root Certificate Authority (Root CA) and multiple Intermediate Credential Authorities. The Root CA is the master root for all other credential authorities (CAs); it is the "center of trust" of the system. The Root

---

[64] The design of DSRC security credentials uses elliptic curve cryptography to sign the credentials. This form of cryptography was chosen because it imposes the smallest possible overhead on the message, leaving the most bandwidth available for safety data.

CA is the only entity that can self-sign a credential–the other CAs cannot. All trust for the system components and users is inherited and delegated from the Root CA through credential issuance.

The basic premise is, just as vehicles and infrastructure in the system need to be "trusted" through the use of short-term credentials that accompany V2V messages, the credential management entities (CMEs) of the SCMS need to be "trusted" by the vehicles or infrastructure as well. The CMEs also need to trust one another. For these reasons, most receive their own digital credentials, referred to as "CME credentials." A vehicle's OBE will examine the CME credential of any digitally signed message it receives before it accepts the message as valid to ensure that the credential has not expired, the CME that issued the credential is trusted, and the credential is not listed on a Credential Revocation List (CRL).

The following elements of the V2V security design are needed for the following reasons:

- The requirement to protect privacy appropriately requires a system that divides and separates some of the functionality to ensure that no one element (entity) has the ability to match records that would lead to identification of a specific driver or specific vehicle.

- There are two linkage authorities (LA) that create linkage values. Linkage values allow one entry on the CRL to revoke an entire batch of credentials, instead of having to list each credential. This drastically reduces the size of the CRL and the communications requirement. An LA has enough information that an inside attacker can track a user. Therefore, the linkage value comes from the output of two separate linkage authorities, neither of which has enough information to track anyone. Splitting the linkage authority creates additional privacy protection but also increases organizational costs.

- The need for appropriate privacy protection has led to a greater amount of digital credential usage; digital credentials use random identifiers that change frequently so as to lower the risk of identifying any one vehicle or driver with a particular digital certificate. The decision on how many credentials are used in a given time period or how to employ random identifiers is still to be determined (options are described but not yet decided upon). It may be a flexible choice based on type of application. Notably, allowing for different schemes might also make attacks on the system more difficult.

- Privacy considerations also have resulted in the addition of an element to obscure location coordinates when a vehicle or device communicates with the system (e.g., to request more digital credentials or to report misbehavior detected locally, around the vehicle).

- While misbehavior authorities exist in today's PKI system (typically as a part of a CA) to detect and take actions to mitigate or remove malicious behavior, the Misbehavior Authority (MA) is described as a separate and more complex entity than exists today. Not all of the described functionality of the MA has been demonstrated (e.g., the use of local detection and reporting) in industry. It is, however, planned for demonstration and testing as an operational prototype that is being planned as part of the ongoing near-term CAMP research.

- The trust requirement has resulted in the design of a direct interface with the SCMS Manager to verify that each type of device meet standards proving their capabilities to be trusted, secure, and interoperable, which will occur through a certification process. [65]

- Request coordination is added as a function to ensure that an OBE cannot obtain multiple batches of credentials by sending requests to several RAs at the same time.

- Last, a device configuration manager is added as a function to provide security and protect privacy when a device is given its initial enrollment credential. This initial credential is the evidence of the device being a trusted player that can then request batches of security credentials that will be sent with the BSM.

3.III.B. System Integrity and Management

In 2013, USDOT sponsored an in-depth organizational study of the proposed SCMS that was designed for V2V communications.[66] One purpose of the research was to generate organizational options for the SCMS by grouping the SCMS functions in into legally, or administratively, distinct entities, in order to enable secure and efficient communications and protect privacy appropriately while minimizing cost. The analysis of the organizational options for the SCMS focused primarily on organizational connections and separations, as well as the closely-related process of characterizing functions as "central" (functions that need to be owned and operated by a single legal entity) or "non-central" (functions that may be owned and operated by multiple legal entities). The issue of whether a function is central or non-central has significant policy implications both for SCMS Organization and for SCMS Ownership/Operation.

The SCMS research began by identifying multiple organizational models that, together, captured all possible configurations of the SCMS functions. USDOT initially selected a small number of these organizational models to address in depth. Ultimately, the organization of the SCMS—the final grouping of functions and estimates of any efficiencies—will be controlled by the organization(s) that manage the SCMS and own and operate the component CMEs. However, NHTSA anticipates being able to influence the organization and operation of the SCMS (and thereby ensure adequate separation to assure secure, privacy appropriate V2V communications) through agreement or Memorandum of Understanding (MOU) with the SCMS Manager or through participation on an SCMS "governance board."

---

[65] As noted previously, there are two certification paths, both of which provide the ability to obtain security credentials. The first path, self-certification, applies to automotive manufacturers who will comply with NHTSA requirements to include DSRC capabilities in their vehicle. The second path is for device manufacturers or application developers to have their products certified through certification laboratories that are now being established to provide such services.
[66] As detailed in the report, *Security Credentials Management System (SCMS) Design and Analysis for the Connected Vehicle System* (Booz Allen Hamilton, Inc., Dec. 27, 2013). [Hereafter, "BAH SCMS Design and Analysis Report"]. See Docket No. NHTSA-2014-0022.

The SCMS Manager is intended to serve as the entity that provides system management, primarily by enforcing and auditing compliance with uniform technical and policy standards and guidance for the SCMS system-wide. The uniform standards/guidance will need to establish and ensure consistency, effectiveness, interoperability, and appropriate security and privacy protection across the CMEs to facilitate necessary communications, sharing of information, and operational connections. The SCMS Manager will need to have mechanisms to ensure that all CME entities have policies, practices, technologies, and communications consistent with system-wide standards and guidance. The SCMS Manager may (but need not) be the body that develops the standards, guidance, or policies applicable system-wide, and would be the entity charged with overseeing standards and policy compliance by the CME entities that, together with the SCMS Manager, make up the SCMS. Technical standards and guidance exist applicable to PKI industry-wide that likely will form the basis for many of the policies and procedures applicable across the SCMS.

3.III.C.   Organization and Ownership

Organizational separation of functions is an example of a policy control often used to mitigate privacy risks in PKI systems—but such separations come with increased costs and may negatively impact the system's ability to identify and revoke the credentials of misbehaving devices.[67] Ultimately, other functions may be co-located within the same SCMS component organization. However, grouping of SCMS functions and any resulting efficiencies or risk trade-offs will depend, in large part, on the system's ownership and operational structure, as well as system governance, and on the preferences of the entity or entities that own and operate the SCMS Manager and CME component entities.

SCMS ownership and operation is inextricably linked to SCMS governance. In essence, there are three basic organizational models that apply *both* to SCMS ownership and operation and to SCMS governance: public, public-private, and private. Due to the lack of Federal funding available to support an SCMS, USDOT research to date has focused on the likelihood of private ownership and operation of the SCMS "industry," with governance being largely "self-governance" by private industry participants and stakeholders, except to the extent that operational requirements may stem from Federal law, regulation, contract or agreement.

As part of its research on the technical framework of a V2V communication security system, USDOT sought stakeholder input on the potential management structure of the communications security system and identified three potential options:

---

[67] *Intelligent Transportation Systems: Vehicle-to-Vehicle Technologies Expected to Offer Safety Benefits, but a Variety of Deployment Challenges Exist.* Located at: http://www.gao.gov/assets/660/658709.pdf, pp.22-23. (Last accessed June 2015)

- *Federal model*: USDOT officials explained that, if the Federal government were to provide the security management services required to support V2V technologies, it most likely would do so through a service contract that would include specific provisions to ensure adequate market access, privacy and security controls, and reporting and continuity of services. According to USDOT, the department has appropriate legal authority to pursue this model but does not have sufficient resources to introduce such a structure at this time.

- *Public-Private model*: Under a public-private partnership, the security system would be jointly owned and managed by the Federal government and private entities. USDOT officials stated that statutory authority would be needed to create a public-private management structure that is vested with the authority that Congress deems necessary and appropriate to finance and operate a V2V communication security system. USDOT officials added that the required legal authority would likely need to include authorization to establish and collect fees on behalf of the entity and possibly provisions addressing liability, privacy, data ownership, and security requirements applicable to such a system.

- *Private model*: USDOT officials stated that its current legal authority and resources have led NHTSA to focus primarily on working with stakeholders to develop a viable private model. USDOT officials suggested that, through an agreement with a privately owned and operated security management provider, a private model could be used to support a V2V communication security system. Aside from any aspects specifically detailed in the agreement, USDOT suggested that the governance and financing of a private management structure would depend on what entity constitutes and owns the entity.

The USDOT's SCMS research indicated that ownership and operation of non-central functions could take different forms. While there are advantages of having different owners (e.g., individual OEMs) oversee large CMEs comprised of all non-central functions, the SCMS research has suggested that running such an overarching CME should not be a *condition* of ownership. Thus, for example, an entity that wants to own and operate one or more Location Obfuscation Proxy (LOP) functions should not necessarily be required to operate *all* of the other non-central functions.

Qualifications for ownership and/or operation of non-central functions may be very different from those required for ownership and/or operation of central functions. For example, due to the critical importance of the security and effectiveness of operation of the Root CA, the owner/operator of this function should have expertise in PKI technology appropriate for the role.

3.III.D.   System Governance

Although heavily dependent on context, the term "governance" generally refers to the way rules are established, implemented, and enforced. Governance can mean formal regulatory oversight by a Federal, State, or local governmental entity. NHTSA's issuance and enforcement of Federal Motor Vehicle Safety System (FMVSS) standard under the Safety Act[68] is an example of governance by a Federal entity. However, governance does not always require the participation of a "government" (i.e., a geo-political entity). In the context of corporate entities, governance typically refers to consistent management, cohesive policies, guidance, processes, and decision-rights for given areas of responsibility.

"System Governance" refers to the body or set of bodies/entities that determine standards, policies, compliance requirements, and expectations for all organizations that have a role to play in certificate management as part of the SCMS that will be needed to support deployment of V2V technologies. It encompasses:

- How decisions are made about various policies, standards, requirements, and practices;
- Who has the authority to mandate and enforce compliance with the policies, standards, and industry requirements;
- Who makes up the overseeing financial, legal, management, and executive operations of the entities in the SCMS;
- How various entities interact with each other;
- How the system addresses privacy issues;
- How risk and liability are allocated across the organizations;
- Who will own the intellectual property (data and software) of the system; and
- How the system's intellectual property will be licensed or otherwise allocated among and between internal and external entities.

The SCMS Manager will define and oversee certain standards, policies, procedures, and operational practices applicable to the SCMS component entities. The potential scope or extent of authority and operations of the SCMS Manager are still under development, but as in all industries, there are three fundamental options for organizational structure from which to choose for SCMS industry governance (the same three apply to the inextricably related issue of SCMS ownership and operation):

- **Public**: denotes a governance structure determined and administered by the government, either directly or indirectly (e.g., via a contractor). The FAA's air traffic

---

[68] As noted in 49 U.S.C. Chapter 301 with a description at: http://www.nhtsa.gov/cars/rules/import/FMVSS/. (Last accessed June 2015)

control system is an example of a direct public governance model. It has a statutory basis, is funded largely by Federal appropriations, and its ownership, control, and operation are subject to Federal laws and procedures. However, absent substantial new appropriations—which NHTSA has no plans to seek at this time—NHTSA lacks the resources to contemplate public ownership, control, or administration of a system the size and scope of the SCMS, as currently conceived.

- **Public-Private Partnership (PPP)**: includes any organizational structure authorized by law within the range of a purely government organization and a purely private organization, established and administered in accordance with the authorizing partnership or similar document (typically a grant, cooperative agreement or other agreement). Under a public-private partnership model, NHTSA would work with the private sector to form a PPP to operate and/or govern the security functions required to support deployment of V2V technologies. Depending on the scope of the agreement, the PPP could be limited to the SCMS Manager functions identified in the current SCMS model, or the PPP could be responsible for owning, financing, and operating the SCMS, as a whole, including the SCMS Manager and CME component entities; or as yet another alternative, the PPP could be limited to forming a governance board of stakeholders to provide input to the SCMS owners/operators. Examples of this kind of governance could be publicly or privately owned utility models, which are complex, highly regulated, and require significant public resources to administer. Due primarily to a lack of current or foreseeable appropriations to support a PPP, USDOT research to date has not fully explored development of a PPP governance model for the SCMS and, instead, has focused on a private model or ownership/operation and governance.

- **Private**: indicates a governance structure established and administered by a purely private organization or organizations, without direct government involvement. A private model could be a viable mechanism for system governance of the SCMS, since NHTSA's existing legal authority will accommodate use of a grant, cooperative agreement, or other agreement to facilitate stakeholder—and even USDOT—input into governance of a private SCMS, assuming willingness on the part of the private entity to enter into such an agreement. Key industry stakeholders have taken a position that a private governance organization, without *any* government involvement (i.e., not under government contract, agreement or MOU), will lack sufficient authority to make all of the decisions and determinations necessary for appropriate system governance of the SCMS. These stakeholders also have expressed other concerns about a purely private governance model, including what it views as lack of stakeholder voice, accountability and government oversight; antitrust risks; potentially increasing costs; and "massive liability exposure."

These governance options have different implications for the level of involvement of the Federal Government and stakeholders in the oversight, setting of policies, rules, standards, procedures, operational practices, liability/risk sharing, funding, and nature of compliance/enforcement within the SCMS industry.

All organizations within the "industry," or all organizations that make up different parts of the SCMS environment, could be represented. The coalition of SCMS "industry" participants, together, could decide on standards, codes of conduct, expectations, and other norms in order to maintain and protect communications security, appropriate user privacy, and required operational functions within the system, under the auspices of the SCMS Manager and/or another coalition-type body. In addition, this group likely would decide on and participate in recommendations about resource management and costs for the industry and its governing body.

Many commercial industries today operate under this model of private governance, establishing private, industry-specific organizations to develop and enforce ethics, standards, code-making, and enforcement functions not specifically required by law. The largest benefit of this kind of governance structure is that it reduces the involvement of the government and therefore reduces the cost to the taxpayers for managing, administering, and enforcing rules within and across the CMEs, although the cost will be passed to the consumer at some point. It also provides more efficiency and flexibility in decision-making than typically is available in the context of a government or PPP model. The positive and negative implications of a private governance structure include:

- Lower costs and more streamlined implementation/operational processes, due to the lack of Federal workplace regulations and processes;

- Need for clear monitoring and enforcement standards and processes, potentially with an additional level of oversight or review/audit; and

- Need for agreements across jurisdictions, organizations, and areas of oversight so as to ensure smooth operations and reduced communications or collaboration challenges.

The private model accommodates some limited Government involvement. Once a coalition or other private entity to serve as SCMS Manager is established voluntarily by a private SCMS "industry," NHTSA could enter into an agreement with that governance entity to ensure that SCMS functions required for V2V safety are delivered by CME entity organizations in a way that is consistent with USDOT's Principles for a Connected Environment. Assuming willingness by the private entity to enter into such an agreement with the Government, either NHTSA or ITS JPO authority might be used to support a mechanism for stakeholder input into SCMS governance.

While the private model possesses considerable benefits, it also carries certain risks that the Federal model does not. The primary risk of a purely private model involves continuity of SCMS function. With no Federal involvement, the party or parties owning and operating the SCMS theoretically could choose to stop doing so at some point. A V2V system needs an SCMS to function; if the SCMS owner/operator ceases to provide the security required for V2V communications, the V2V system will no longer work. Even with some amount of Federal involvement, it remains difficult to compel specific performance if the performing party chooses to stop performing. One option for minimizing the not insignificant risk associated with a private model, should NHTSA enter into an agreement with a private SCMS owners/operator, is to include certain contractual provisions in the agreement. NHTSA can structure the agreement so that the private SCMS owners/operator is required to provide sufficient notice of its intent to cease providing V2V security services and to continue operating the SCMS until NHTSA can identify another entity to assume operations, or so that the Federal Government receives liquidated damages in the event of non-performance. Of course, operational disruption also could be a risk under a Federal model if Congress suddenly withdraws funding after NHTSA establishes the SCMS. In any event, a thorough consideration of contingencies for risks such as this seems highly advisable.

3.IV. Essential Standards

The Institute for Electrical and Electronics Engineers (IEEE) 802.11 Wi-Fi standard describes the performance of DSRC. Only certain parts of this industry standard are required for implementing DSRC operations. To accommodate the rapid exchange of trajectory information between vehicles traveling at high speed, IEEE 802.11 was amended—802.11p— to enable operation without setting up a basic service set.[69] This amendment further:

- Allows security services, such as authentication, to be provided by other standards; and
- Describes adjacent channel and alternate adjacent channel interference criteria and transmission masks corresponding to requirements of the FCC rules for DSRC.

Additional standards are required to ensure interoperability among various makes and models of devices and communications technologies, and to ensure that messages are of appropriate quality and are trusted and authenticable. The essential cooperative system standards that facilitate DSRC operation are:

- **IEEE 802.11p** to establish a wireless link for V2V and V2I communications (the entire standard applies to V2V and V2I communications, because it defines the structure for how devices should communicate using the 5.9 GHz frequency band);
- **IEEE 1609.x** to establish protocols for information exchange across the wireless link; and
- **SAE J2735 and SAE J2945.x** to define message content for communicating specific information to and from equipment and devices via DSRC or other communications technologies.

Appendix E.II defines each of these standards in greater detail and provides their current status. In summary, the critical standards are stable; the published versions were employed to develop the applications and devices for the Safety Pilot Model Deployment test bed. Using lessons

---

[69] A definition for a basic service set can be found here: http://www.juniper.net/documentation/en_US/network-director1.5/topics/concept/wireless-ssid-bssid-essid.html (Last accessed June 2015)

learned, the standards working groups are working on the next versions for publication in the near future.

The ITS JPO's Standards Program funds and manages ITS cooperative system standards efforts and international harmonization (harmonization efforts are discussed in Appendix H). The content of these standards is developed collaboratively with contributions from diverse stakeholders. The CAMP research has resulted in significant contributions to many of the standards described above.[70]

3.V. DSRC Maturity and Status

3.V.A. Status and Maturity of the DSRC Connected Vehicle Technologies and Applications

A decade of research and development advancements on DSRC technologies, channel testing, applications, and standards provide a near comprehensive set of results as a basis for deployment. USDOT and stakeholders such as CAMP, the VIIC, Vehicle Safety Consortium (VSC, a subset of CAMP members who participated in specific V2V studies), AASHTO, State and local agencies, and universities have conducted analyses; prototyped and tested applications, technologies, security solutions, and new institutional arrangements. These stakeholders also have gathered findings and lessons learned to incorporate into final versions of performance specifications, architecture viewpoints, and standards. In parallel, many test beds and pilot deployments have put DSRC into use for research purposes and are preparing plans to transition to operational uses.

The assessment of USDOT and stakeholders is that DSRC technologies and applications have reached a level of stability that supports deployment. This is evidenced by the following actions by USDOT:

- **Deployment Preparations:** Large-scale testing in the Safety Pilot Model Deployment that has resulted in the FY2015 plans to launch Connected Vehicle Pilot Deployments:
  - Devices tested in the Model Deployment were developed based on existing communication protocols found in voluntary consensus standards from SAE, IEEE, NHTSA, and others participating in the Model Deployment. With findings that the standards were open to much interpretation, standards working groups (including USDOT support) have been developing additional protocols that

---

[70] Specifically, CAMP VSC have contributed to the development of SAE J2735 (DSRC Message Set Dictionary); SAE J2945.1 (DSRC BSM Minimum Performance Requirements); IEEE 1609.0 (Architecture); IEEE 1609.2 (Security Services); IEEE 1609.3 (Networking Services); IEEE 1609.4 (Multi-Channel Operation); IEEE 1609.12 (Identifier Allocations); and IEEE 802.11p (Wireless Access in Vehicular Environments (WAVE)).

assure device interoperability. These updates ensure that the standards will support a larger, widespread technology rollout.[71]

o Hardware has been evolving from the pre-competitive, prototype components used in the Model Deployment into products for full-scale production. NHTSA's current understanding, based on discussions with industry OEMs and suppliers, is that securing and preparing manufacturing facilities is the major factor to transitioning from building prototype components to ramping up to produce mass-market components, and that DSRC devices in their current form are nearly production-feasible today.[72]

- **Policy Development:**
  - o **NHTSA Decision to move toward Rulemaking for Light Vehicles**
    - In 2014, NHTSA began work on a regulatory proposal that would require V2V devices in new vehicles in a future year, consistent with applicable legal requirements, Executive Orders, and guidance. USDOT believes that the signal this announcement sends to the market will significantly enhance development of this technology and pave the way for market penetration of V2V safety applications. NHTSA has announced its intention to complete a V2V NPRM in early 2016.
  - o **FHWA Guidance Development**
    - Following on the NHTSA announcement, FHWA is engaging with stakeholders on their needs for Guidance and tools to support deployment.
    - Draft Guidance is being vetted by the Department and will be released in late 2015.
    - In Summer 2014, FHWA launched additional analysis needed to produce the tools and reference materials to support the Guidance including:
      - A DSRC Licensing Guide;
      - A Connected Vehicle Planning Guide;
      - A Connected Vehicle Systems Engineering Guide;
      - A Connected Vehicle Deployment Guide;
      - An analysis on the extensibility of the V2V security solution to address the service needs for V2I, V2X and other points of connection; and

---

[71] NHTSA: *Vehicle-to-Vehicle Communications: Readiness of V2V Technology for Application*, DOT HS 812 014, pg. 15, at: http://www.nhtsa.gov/staticfiles/rulemaking/pdf/V2V/Readiness-of-V2V-Technology-for-Application-812014.pdf (Last accessed June 2015)
[72] Ibid, page 87.

- An analysis of the legislative challenges and opportunities to Connected Vehicle deployment in the States.

- **Tool Development:**
  - **Connected Vehicle Reference Implementation Architecture and the Systems Engineering Tool for Intelligent Transportation (SET-IT)**
    - The alpha version of the CVRIA has been reviewed and commented on by stakeholders, ensuring that the tool is evolving toward maturity. Webinars on the CVRIA held in 2013 attracted nearly 1,000 participants across the nation and the world. Key stakeholders such as the VIIC and AASHTO have launched efforts to provide detailed reviews on the viewpoints that impact them.
    - In 2014, USDOT launched a tool—the SET-IT tool—that integrates both drawing and database capabilities with CVRIA and allows users to develop project architectures for CV pilots, test beds, and early deployments that are tailored to their specific location and needs. [73]
  - **Development of a Connected Vehicle Standards Plan and International Harmonization**
    - The plan will guide the completion of critical Connected Vehicle standards and identify priorities of USDOT and State and local agencies as they prepare for deployments and Connected Vehicle Pilots.
    - USDOT has also invested in the Southeast Michigan 2014 implementation project, which is developing new and dynamic data warehousing concepts and Ethernet frame messaging protocols that ensure messages will be "communications-agnostic" within a Connected Vehicle environment.
  - **Sensor Reliability and Integration/Interoperability**
    - NHTSA intends to require system reliability requirements by regulating malfunction-reporting requirements for V2V systems. For example, an OBE needs to integrate a GPS receiver and on-vehicle sensors including accelerometer, steering wheel angle sensor, brake system sensor, and yaw sensor to provide the BSM data elements. The accuracy and reliability of GPS and sensor data are critical to V2V and V2I applications. Accuracy requirements of sensor data will be specified in industry consensus standards (SAE J2735 and SAE J2945) and in NHTSA's NPRM. These standards are expected to specify absolute geo-

---

[73] The tool can be found at: http://www.its.dot.gov/arch/set_it.htm. (Last accessed June 2015)

location accuracy, speed and yaw rate sensing accuracy, timing requirements, and other accuracy requirements for data elements mandated for the BSM. GPS availability measurement and testing are outlined in detail the in CAMP's Vehicle Safety Communications– Applications (VSC-A) Final Report.[74]

▪ RSEs need to integrate the DSRC module with traffic signal controllers or other traffic control devices. Essentially, the DSRC module may retrieve Signal Phase and Timing (SPaT) data from a traffic signal controller through an Ethernet port (or a serial port) using National Transportation Communications for ITS Protocol (NTCIP). The accuracy of SPaT and intersection map data can be verified by checking with the SPaT setting of traffic signal controllers. Due to the hardware constraint of traffic signal controllers (only new traffic signal controllers, e.g., NEMA TS2 and 2070 may have equipped with NTCIP and the communication interface), the certification of RSE requires an integration prerequisite of the DSRC module and traffic signal controller.

▪ In the Safety Model Pilot Development conducted from 2012 to 2014 in Ann Arbor, Michigan, state-of-health monitoring was conducted for both vehicle-based devices and infrastructure-based devices. The test results showed that most malfunctions were related to software problems and only few hardware glitches (power cable connector replaced and battery drained) were encountered.[75] The accuracy and reliability of OBEs and RSEs can be warranted by NHTSA's minimum performance requirements including automotive-grade hardware specifications.

- **Security and Privacy—A Security Credential Management System:**
    - o USDOT, CAMP, and security experts have developed an initial security and privacy protection solution that will be prototyped and tested in 2016. It is expected to service initial vehicle fleets, infrastructure components, and applications.

---

[74] *Vehicle Safety Communications – Applications (VSC-A) Final Report,* September 2011, DOT HS 811 492A, http://www.nhtsa.gov/DOT/NHTSA/NVS/Crash%20Avoidance/Technical%20Publications/2011/811492A.pdf, last accessed May, 2015.

[75] *The Ann Arbor Safety Pilot Model Deployment Experience – Task 7 Appendix A,* a Safety Model Pilot Development report submitted to NHTSA by the University of Michigan Transportation Research Institute and Partner Organizations, 2014.

- **Certification:**
    - In 2015, USDOT awarded three laboratories grants to work cooperatively with industry to develop specialized test equipment; new certification processes and test procedures; and institutional arrangements to support certification for connected vehicle activities as we move toward nationwide deployment.

3.V.B. Status and Maturity of DSRC Implementation

Successful use of the 5.9 GHz DSRC band is governed by service rules that are documented in the FCC's Report and Order (R&O) of 2004, and amended in 2006. Using these rules, USDOT and its industry partners, test partners, test bed operators, and stakeholders have performed rigorous research, analysis, and development and testing on DSRC technologies, applications, and standards. The results of these efforts provides a near-comprehensive set of results as a basis for DSRC use. Technology and applications development have led to testing of prototypes to confirm how they operate according to the FCC rules. Certification test procedures for communications will be based on these rules. Appendix I describes the rules and how they guide operations of the DSRC band of the spectrum.

Scalability is one of the critical factors that suggest whether DSRC operations are ready for nationwide implementation. NHTSA and CAMP research results show that the V2V safety applications perform reliably in test scenarios with up to 1200 vehicles (using simulation) in communication range.[76] Ongoing research in 2015 is being structured to estimate fully the number of other DSRC-equipped vehicles that a single DSRC radio would be exposed to in an environment (such as heavy freeway traffic) where channel congestion could be significant.[77]

CAMP is leading the work for NHTSA related to evaluating scalability challenges and developing potential solutions. CAMP has recently delivered to NHTSA two pivotal draft reports summarizing scalability testing to date, and potential solutions. Specifically, CAMP has developed standardized algorithms for detecting bandwidth limiting conditions that may develop in certain traffic congestion conditions, and subsequently implementing mitigation measures such as reducing transmission power and/or transmissions per second. Work thus far suggests the congestion mitigation algorithms will be sufficient to manage scalability concerns while having zero or minimal impact on the safety performance of the applications. Work will continue on refining the congestion mitigation algorithms. USDOT expects to work with CAMP on

---

[76] CAMP had conducted a series of channel congestion tests using 400 OBEs to emulate 400 connected vehicles. The tests of 2000 connected vehicles were implemented with the simulation software NS3 by increasing the transmission frequency to 5 times (50 Hz) of the BSM transmission rate (10 Hz).

[77] NHTSA: *Vehicle-to-Vehicle Communications: Readiness of V2V Technology for Application*, DOT HS 812 014, pg. 125, at: http://www.nhtsa.gov/staticfiles/rulemaking/pdf/V2V/Readiness-of-V2V-Technology-for-Application-812014.pdf (Last accessed June 2015)

finalizing acceptable, standard scalability solutions in late 2015. The two CAMP Scalability Reports are expected to be placed in the public docket in the coming months.[78]

Misbehavior Detection (MBD) is an essential component of an overall security solution. The goal of this current activity is to introduce potential methods of addressing misbehavior that are expected to occur within a V2V network. To do so, MBD was separated into two parts, namely local and global MBD. Using industry-accepted threat, vulnerability, and risk analysis techniques, a list of potential attacks was developed[79] and further narrowed to a describe subset of attacks deemed highly relevant to current V2V safety applications. The attack analysis was used as input in developing a number of potential local and global MBD methods. A preliminary list of data elements was also identified as needed for the interface between the local MBD methods at the vehicle level and global MBD methods of the Misbehavior Authority inside of the SCMS.[80]

Another implementation issue being addressed is human factors, and, in particular, safety alert prioritization and driver overload. Building from earlier studies, NHTSA is currently in the process of finalizing a comprehensive synthesis study titled: "*Driver Vehicle Interface (DVI) Design Assistance for Advanced Vehicle Technologies*". This report will offer design principles useful in developing driver warning and notification interface systems for advanced crash avoidance systems—including those based on V2V and V2I communications. This study also addresses, among other DVI design considerations, prioritization of driver warnings and alerts as well as driver workload and human–systems integration issues. NHTSA intends to place this report in the public docket in the coming months of 2015.

---

[78] CAMP, *Interoperability Issues of Vehicle-to-Vehicle Based Safety Systems Project (V2V-Interoperability) Phase 2 Final Report Volume 1 – Communications Scalability for V2V Safety Development*, Feb., 2015. CAMP, *Interoperability Issues of Vehicle-to-Vehicle Based Safety Systems Project (V2V-Interoperability) Phase 2 Final Report Volume 2 – Communications Scalability for V2V Safety Analysis*, March, 2015.

[79] The attack scenarios are replay, tunnel, framing, DoS (denial of service) at physical layer, DoS at application layer, DoS by physical tampering, authenticated malformed message, internal sensor spoofing, external sensor spoofing, security credentials extraction, non-authenticated messages, slander, and Sybil attack. CAMP, Interoperability Issues of Vehicle-to-Vehicle Based Safety Systems Project (V2V-Interoperability) Phase 2 Final Report Volume 3 – Security Research for Misbehavior Detection, will be published in late 2015.

[80] SCMS operations consist of system initialization, system management, certificate addition, certificate removal, certificate rollover, certificate revocation, device certification, key management, misbehavior detection, and reinstatement.

The ITS JPO and NHTSA have performed past research on safety alert prioritizations for heavy vehicles.[81] The human factors research for DSRC-based safety systems has built on the lessons learned in the development and market adoption of these systems.

In summary, the research and development activities have addressed most of the challenges identified to DSRC use with Connected Vehicle implementation.

3.V.C.   Analysis of Known and Potential Gaps

As discussed previously, MAP-21 Section 53006 (a) requires, in part, that the DSRC report analyzes known and potential gaps in short-range communications technology. DSRC technology is sufficiently mature that gaps have predominantly been addressed. What remains is the work to provide greater specificity on issues such as spectrum usage, performance requirements, or final standards.

Recently, two issues have arisen that are providing uncertainty for vendors making investments and early adopters formulating plans. From a technical perspective, the first issue is the request to consider coexistence with unlicensed devices. Implementers need the assurance that possible sharing of the band of radio-frequency spectrum used by V2V and V2I communications will not jeopardize their crash avoidance capabilities. These unlicensed Wi-Fi devices (referred to as Unlicensed-National Information Infrastructure, or U-NII) wireless broadband devices are expected to become widespread in use in the foreseeable future. It is important to ensure that changes to the 5.9 GHz DSRC band do not jeopardize crash avoidance capabilities. These U-NII devices have yet to be tested to determine if they will interfere with crash-avoidance applications or results in unacceptable risks to traveler safety.

Second, from an organizational perspective, the emerging issue is the possibility that interference with the 5.9 GHz band could negatively impact the delivery of communications that form the basis for safety-critical warnings to drivers. If this were to happen, active frequency coordination might be required to address the problem.

In addition to these two issues, there exist a small set of open items that need further specificity to support implementers. Many of these items were identified as gaps by the NRC although they are addressed in the NHTSA Technology Readiness report that accompanied the V2V ANPRM.

---

[81] Known as the Integrated Vehicle-Based Safety System project, the results produced an integrated countermeasure system to address rear-end, run-off-road, and lane change crashes. See more at: http://www.its.dot.gov/ivbss/. (Last accessed June 2015)

3.V.C.1. Coexistence With Unlicensed Users

On June 28, 2010, President Obama issued a Memorandum directing the Secretary of Commerce to work with the FCC to identify and make available 500 megahertz of spectrum over the next ten years for wireless broadband use. The FCC's Policy and Plans Steering Group (PPSG) comprised of Federal agency members (including the Department of Defense) recommended adding the 5350-5470 MHz and 5825-5925 MHz bands to the bands under consideration, the latter of which overlaps with the DSRC/ITS radio spectrum. On February 22, 2012, the President signed the Middle Class Tax Relief and Job Creation Act of 2012 into law. The Act requires the Assistant Secretary of Commerce (through NTIA), in consultation with the Department of Defense (DoD) and other impacted agencies, to evaluate spectrum-sharing technologies and the risk to users if Unlicensed-National Information Infrastructure (U-NII) wireless broadband devices were allowed to operate in these bands.

The most common example of U-NII devices is a Wi-Fi device that operates without a license in a specific band, and which has no interference protection. These devices are expressly prohibited by the FCC from interfering with licensed devices operating within their authorized bands. If these types of U-NII devices are allowed to operate in the 5.9 GHz DSRC band and they result in harmful interference, it may be impossible to remove them. Interference could be caused by:

- Incomplete testing;
- Power or noise levels that preclude DSRC devices from functioning;
- Unlicensed devices implementing a detect and vacate function that does not protect DSRC operations;
- Poor quality control in device manufacture; or
- Intentional violations of the spectrum protocols.

With regard to the last bullet, research has documented the problems that FAA has faced with interference from unlicensed devices. This experience has revealed that, once introduced, unlicensed use is not only a challenge to curtail but that tens of millions of taxpayer dollars have been spent trying to eradicate these problems in the name of safety.[82]

USDOT finds that the emerging industry spectrum sharing proposals do not yet offer the detail needed to evaluate whether they would interfere with DSRC technologies and applications.

---

[82] Documentation of the most recent details can be found in the NTIA Report TR-12-486: *Case Study: Investigation of Interference into 5 GHz Weather Radars from Unlicensed National Information Infrastructure Devices, Part III,* located at file: www.its.bldrdoc.gov/publications/download/12-486.pdf. (last accessed November 2015)

Because no devices exist at this time, little is known about how U-NII devices would operate within the DSRC/ITS Services band of the spectrum. Early USDOT analyses indicate that U-NII devices would deny or interfere with safety-critical V2V messages. A recent set of experiments conducted by Korea University conclude that *"If [wireless local area network] WLAN devices get to share the 5.9 GHz DSRC band as the FCC plans, they can undermine the safety-critical vehicular communication…the WLAN technology in this band is expected to use a shorter symbol time and smaller inter-frame space, so they will have relative edge in channel use…"*. [83]

In recent months, USDOT's efforts related to evaluating spectrum sharing have expanded in several important ways that will enable the Department to evaluate risks thoroughly to the DSRC/ITS services from both unlicensed and licensed devices operating in the 5.9 GHz band.

First, building on initial research that was done in the late 1990s and in the 2000s, when the suitability of the 5.9 GHz band for DSRC applications was originally evaluated, USDOT engaged spectrum experts who performed tests and analyses to measure noise and interference from existing sources in the band and in adjacent bands. These include fixed satellite services, microwave links, and military radar applications. [84] These data helped USDOT and FCC better understand existing conditions. These data now form a baseline against which to compare proposed additional licensed and non-licensed uses of the 5.9 GHz band. In particular, baseline data will help in evaluating the implications of changing the designated band for BSM transmission from its current channel 172 to a higher channel (180 or 182) as is being proposed by some industry stakeholders (since the spectrum interference profile will be different for these different bands).

USDOT is also implementing a comprehensive test plan related to spectrum sharing that will evaluate the two leading proposals being offered by the unlicensed device industry (specifically, "detect and vacate" and "re-channelization" proposals), and will include other sharing approaches that may offer safe-sharing while still protecting the safety aspects of DSRC. USDOT has begun testing and is actively seeking industry partners to provide unlicensed devices for testing with DSRC devices. At this time, it is anticipated that results will be available by December 2016 (dependent upon various factors, including when devices are provided).

---

[83] *On the Coexistence of IEEE 802.11ac and WAVE in the 5.9 GHz Band* by Yongtae Park and Hyogon Kim, Korea University. Published in IEEE Communications Magazine, June 2014, pages. 162-168.

[84] References include a report by Johns Hopkins, located at:
http://apps.fcc.gov/ecfs/document/view;jsessionid=1kvXQdfN3ysprYvz3GmkCfTwXStrbDhCh51nc1WhyLBSpdjGBbRd!-856245186!973241960?id=6516718563; a report by NTIA and the Department of Commerce, located at:
http://www.ntia.doc.gov/files/ntia/publications/00-373.pdf; and information summarized in the FCC's notice of proposed rulemaking (ET Docket No. 98-95), located at:
https://transition.fcc.gov/Bureaus/Engineering_Technology/Notices/1998/fcc98119.pdf. (Last accessed June 2015)

USDOT is additionally building and tailoring a bandwidth utilization model, and estimating bandwidth needs from emerging V2I, V2X, and automated vehicle applications in an effort to better define how these applications may utilize the available 75 MHz of spectrum allocated by FCC for vehicular use. USDOT is working in partnership with the FCC and NTIA to define these test plans and models; USDOT will provide NTIA with baseline, bench, and field test data/results for further simulation and modeling to identify impacts when DSRC and other uses of the 5.9 GHz band are scaled to a National level.

The FCC and NTIA regulate the use of spectrum for private and government purposes, respectively. USDOT engages with FCC and NTIA in a variety of ways to examine spectrum use and to determine ways in which spectrum may be used more efficiently, including the potential for spectrum "sharing," where sharing can be accomplished consistent with law and public policy. USDOT supports the President's commitment to make available a total of 500 MHz of Federal and non-Federal spectrum over the next 10 years, suitable for both mobile and fixed wireless broadband use. However, USDOT must also ensure that the safety purposes for which spectrum has already been allocated are not compromised.

More specifically, USDOT is engaged in discussions with FCC, NTIA and other stakeholders on the spectrum allocated for the purposes of the Connected Vehicle program, including the potential "sharing" of spectrum to be used in the 5.9 GHz band. USDOT believes that FCC and NTIA must ensure that unlicensed devices do not compromise safety through harmful interference to the ITS architecture, operations, or safety critical applications if permitted to operate in that band. We have serious concerns about any spectrum sharing that prevents or delays access to the desired channel, or otherwise pre-empts the safety applications.

USDOT is working diligently in concert with FCC, NTIA, other partners to evaluate whether it is feasible to share the spectrum allocated for vehicle safety with unlicensed wireless devices, without those devices interfering with DSRC communications in ways that are detrimental for safety.

Path Toward Resolution of Coexistence with U-NII Devices
USDOT is performing analyses and developing test procedures to evaluate coexistence with unlicensed wireless broadband users. USDOT began with participation in the IEEE 802.11 Tiger Team weekly discussions to understand the proposed characteristics of the unlicensed devices.

USDOT and spectrum experts have also been preparing test sites and collecting data on the existing noise and interference. These measurements will offer a baseline for when USDOT tests the concept of coexistence. For this testing, USDOT has invited industry to offer their devices; if no devices are offered, they may be developed by USDOT and contracted experts to mimic the known characteristics of U-NII devices. Data from small-scale tests will be provided to NTIA for simulation and modeling to determine how such devices may perform on a nationwide scale.

Further, industry organizations that are interested in unlicensed use have been invited to participate in the Southeast Michigan 2014 project and to bring devices that can be tested in an operational setting. USDOT is also working with CAMP and other experts to define harmful interference so that crash avoidance safety applications can be tested in the presence of U-NII devices.

3.V.C.2. Frequency Coordination

As DSRC technology has developed and the rules governing its use have solidified, stakeholders and experts are highlighting a potential institutional gap associated with a desire for a more active form of frequency coordination. Two underlying concerns are:

- The potential for interference amongst proximate, unrelated roadside DSRC units (RSUs); and

- The potential need to address interference associated with use of unlicensed devices. FAA experience with the TDWR systems provides a reference point for this potential need.

To mitigate these concerns, technical spectrum coordination techniques and organizational management roles have been under discussion. Although the FCC declined to establish such an approach when requested in 2004[85], the requirements for using spectrum associated with safety-critical applications are now better understood.

USDOT is performing research to define the needs associated with frequency management, and to clearly and concisely propose the coordination that would be needed with FCC and NTIA. Key research questions include the following:

- What coordination is necessary to ensure that both public and private users gain access to the spectrum appropriately, while maintaining priority for safety?

- What needs will agencies and deployers have for a frequency coordinator, particularly in the early stages with minimal infrastructure?

- What are the characteristics of an organization that can maintain an objective and independent perspective to set and maintain appropriate standards as technologies evolve? Are there roles for more than one organization?

- How will the coordinating organization financially sustain its role?

Path Toward Resolution of Frequency Coordination

More work is needed to clearly and concisely describe the frequency coordination issue, its potential impact on safety, and its possible effects on DSRC deployments around the country if not effectively addressed. USDOT will work with stakeholders, including the FCC and NTIA, to

---

[85]http://apps.fcc.gov/ecfs/document/view;jsessionid=sK3cQdpGLCGPDp1bQCdybsRVbYjkqbc1lLDjkL9yn93XlypvyLh Q!973241960!-856245186?id=6516483012 (Last accessed June 2015)

identify characteristics needed for entities to become spectrum managers, including financial sustainability, as well as propose roles.

3.V.C.3. Open Items

Finally, a few open items will need to be addressed to support finalization of the policy foundation and deployment:

1. Deciding whether to leave the ASTM[86] International Standard in the FCC R&O, or to have the FCC reflect the current IEEE standard

2. Providing the FCC with certification test procedures and supporting documents

3. Communicating the USDOT position on liability

ASTM Standard versus IEEE Standard
In 1999, as the FCC was preparing the first Report and Order to govern the 5.9 GHz DSRC band, the wireless performance requirements were described in an ASTM E2213-03 standard. As these requirements evolved, the standards work migrated to the IEEE 802.11 working group to be addressed as part of the standard that defines Wi-Fi requirements, resulting in the 802.11p amendment which, to date, has been kept consistent with the ASTM document. However, all new enhancements are being captured in the IEEE 802.11 standard.

The open item is a question as to whether to change the language in the FCC rule to reflect this shift. Further analysis to identify appropriate steps to deal with this item is underway with ITS America and AASHTO. USDOT will coordinate any further steps with FCC, NTIA, and other stakeholders as appropriate.

Certification Test Procedures
Certification test procedures exist in draft form and will be demonstrated using existing laboratory and/or other partners' testbeds. The lessons learned from this testing will inform the completion of certification procedures in mid-2016. Section 4.II.E and Appendix G provide detail about the proposed certification path and status of the research to define certification test procedures.

Liability
The NRC noted liability as an open item. However, NHTSA's Technology Readiness report addressed the issue of liability, noting that because V2V technologies under consideration by NHTSA involve alerts and warnings only, USDOT does not believe they present increased or

---

[86] Formerly known as the American Society for Testing and Materials.

---

unbounded liability exposure for the auto industry. [87] Similarly, the V2I applications being researched by USDOT (and which may be offered by local roadway operators, or by third party service providers) also are focused on providing the driver with warnings and/or increased situational awareness only (e.g., "work zone ahead"; "low bridge ahead"; etc.).

Also discussed in NHTSA's Technology Readiness report, under the existing product liability tort framework, manufacturers can limit their legal liability effectively by providing consumers with adequate warnings and instructions for using V2V equipment. Such consumer warnings and instructions typically emphasize the limited role of safety warning technology and explain the limitations of the systems. Such liability considerations are specifically noted in the Dykema Risk Assessment Report referenced in NHTSA's Technology Readiness report. Newer vehicle models currently on the market that are equipped with systems such as lane-departure warning, back-over detection warnings, and forward vehicle detection typically follow this approach in carefully describing the operation and limitation of these systems.

USDOT expects that the OEMs will follow this same approach to limiting their potential liability in connection with V2I safety, mobility and environmental applications – and that OEMs explicitly will disclaim liability for damages or injuries stemming from the use of third party V2I services or applications.

There are potential liability risks to various infrastructure owners and operators from use of new V2I technologies. Parties that could potentially face liability from failures of V2I technology include state and local governments, railroads, bridge owners, and roadway owners and operators. As evidenced by the numerous lawsuits claiming that failure of a traffic signal contributed to an accident, such cases often are brought against public or quasi-public entities. Liability law and precedent related to such failures are well established, and V2I applications and technology are likely to represent an extension of existing roadway infrastructure liability law. NHTSA's Technology Readiness report outlines the various mechanisms the Federal government has at its disposal to limit the liability of private and public entities and individuals, when Congress deems it appropriate to do so. These mechanisms, which are described in the NHTSA report, include: explicit and implicit preemption of liability, contractual indemnification, statutory immunity, capped liability, risk transfer, insurance pools and reinsurance programs.

---

[87] NHTSA: *Vehicle-to-Vehicle Communications: Readiness of V2V Technology for Application*, p. 208, DOT HS 812 014, at: http://www.nhtsa.gov/staticfiles/rulemaking/pdf/V2V/Readiness-of-V2V-Technology-for-Application-812014.pdf. (Last accessed June 2015)

3.V.D   GAO Report to Congress Assessment on V2V and DSRC Readiness[88]

In 2013, the GAO reviewed documentation on V2V technologies and the research efforts, visited the Safety Pilot Model Deployment, and interviewed stakeholders to understand the status of V2V technologies. GAO examined:

- The state of development of V2V technologies and their anticipated benefits;
- The challenges, if any, that will affect the deployment of these technologies and what actions, if any, USDOT is taking to address them; and
- What is known about the potential costs associated with these technologies.

GAO found that the development of the technologies had progressed to the point of real world testing; if broadly deployed, the technologies would offer significant safety benefits.

The report identifies the remaining challenges to V2V deployment and notes that USDOT is working to address these challenges. Stakeholder[89] concerns revealed by GAO embody one of the identified challenges, namely, that any possible sharing of the band of radio-frequency spectrum used by V2V communications should not jeopardize crash avoidance capabilities. GAO noted that the NTIA completed a qualitative evaluation in January 2013[90] regarding spectrum-sharing technologies and the risks to Federal users associated with allowing unlicensed devices to share the 5.9 GHz DSRC band. NTIA concluded that further work was needed to determine whether and how the risks identified can be mitigated.

GAO further noted that sharing the band may create an added burden for both automobile manufacturers and suppliers who would have to consider technical steps to make coexistence with unlicensed devices feasible. To determine if such a burden exists, all parties would need to conduct additional testing to maintain confidence that V2V technologies will work as envisioned.

USDOT has coordinated with NTIA to submit its concerns, through comments to the FCC, that sharing the allocation could degrade the performance of V2V safety applications. As NTIA continues its analysis of potential risk mitigation strategies, USDOT is working cooperatively with NTIA to examine proposed spectrum-sharing arrangements.

---

[88] The following section is summarized from: *Intelligent Transportation Systems: Vehicle-to-Vehicle Technologies Expected to Offer Safety Benefits, but a Variety of Deployment Challenges Exist.* Located at: http://www.gao.gov/assets/660/658709.pdf. (Last accessed June 2015)

[89] Included were four automobile manufacturers and 16 experts.

[90] Located at: http://www.ntia.doc.gov/files/ntia/publications/ntia_5_ghz_report_01-25-2013.pdf. (Last accessed June 2015)

3.V.E   Relationship of DSRC to Emerging Automated Vehicle Technologies

USDOT envisions that DSRC connectivity will also play a major role in the development and deployment of automated vehicles, with V2V communications acting as a foundation for continuing to ensure safety. Automated vehicles may use on-board sensors, cameras, GPS, and telecommunications (i.e., DSRC) to obtain information in order to make their own judgments regarding safety-critical situations and act appropriately by effectuating control at some level. While vehicles equipped with only V2V technology are not automated vehicles, the warnings generated can provide valuable information to augment active on-board safety control technologies and automation systems. In fact, the realization of the full potential benefits and broad-scale implementation of the highest level of automation may conceivably rely on V2V technology as an important input to ensure that the vehicle has full awareness of its surroundings.[91]

USDOT finds the Connected Vehicle and Automated Vehicle research areas as complementary, even necessarily intertwined, and not mutually exclusive areas of focus. The Department expects that Connected Vehicle technologies and applications will be incorporated in Automated Vehicles as they emerge in the marketplace. Notably, the crash avoidance systems available on some cars today (and that will be essential to any "hands-free" driving) could be greatly enhanced by data provided by Connected Vehicles. V2V and V2I communications will enable automated functions to perform better. Even 'self-driving' cars will need to be aware of the vehicles around them, to communicate with traffic signals, and to sense when pedestrians, bicyclists, and motorcycles are in danger. For this to happen in a reliable and cost effective way, vehicles will need wireless technology that gives them 360 degree awareness of other vehicles and the surrounding infrastructure, even when out of sight.

---

[91] *National Highway Traffic Safety Administration Preliminary Statement of Policy Concerning Automated Vehicles*, located at: http://www.nhtsa.gov/staticfiles/rulemaking/pdf/Automated_Vehicles_Policy.pdf. (Last access June 2015)

Chapter 4 DSRC Path for Implementation

This chapter describes a more detailed and recommended implementation path for DSRC inclusive of such tools as the National ITS Architecture, ITS Standards, certification, and policies/guidance. It addresses Congress's requirement for "not preferencing the use of any particular frequency for vehicle-to-infrastructure operations." This chapter describes an emerging, envisioned implementation path for Connected Vehicle environments. It is now possible to articulate a roadmap using a synthesis of USDOT and stakeholder research results.

4.I. Implementation of Connected Vehicle Environments

USDOT and other Connected Vehicle stakeholders, including AASHTO and the OEMs, are developing visions, plans, and guidelines for how the Connected Vehicle environment may be implemented. Those visions and plans explicitly assume and depend upon use of DSRC for safety-critical applications. The AASHTO Footprint Analysis was led by their Connected Vehicle Deployment Coalition, composed of representatives of State departments of transportation, including those that have already installed and are operating DSRC equipment on public roads. The OEMs have worked with NHTSA to identify expected penetration rates of vehicles that are manufactured with integrated devices. Separately, USDOT has considered various scenarios for aftermarket device penetration.

Working with stakeholders, USDOT has synthesized their plans into a candidate roadmap to guide industry and public agency implementation efforts. The preliminary roadmap anticipates advancements in implementation during three overlapping timeframes—2015-2025, 2020-2035, 2030-2040—during which the preponderance of communications capabilities within the vehicle fleet will evolve. The roadmap is grounded in results of stakeholder research and analysis, and offers three key milestones for the year 2040:

Milestone	Description
Vehicles	⇒ 90% of the national light vehicle fleet could include DSRC, allowing the vehicles to communicate with one another and with roadside infrastructure. This percentage assumes that USDOT pursues rulemaking for deployment of DSRC on-board equipment in light vehicles to enable crash avoidance safety applications.[92] The milestone is also based on a range of vehicle penetration with market sales envisioned as starting in 2020.
Infrastructure	⇒ 80% of all traffic signals could be DSRC equipped, or approximately 250,000 signals in total.[93] This milestone for traffic signal instrumentation may be adjusted once State and local planning analyses are conducted; but assumes an approximately 20-year signal equipment replacement cycle and phase-in of DSRC equipment as older equipment is retired.
Expansion	⇒ Up to 25,000 additional roadway locations could be DSRC equipped.[94] This milestone assumes an initial targeting of existing ITS infrastructure sites—locations where Connected Vehicle technologies and applications may augment or replace existing traveler information dissemination and/or data collection equipment. It also incorporates the AASHTO Footprint DSRC RSU deployment projections.

Image Source: USDOT/AASHTO report

[92] NHTSA: *Vehicle-to-Vehicle Communications: Readiness of V2V Technology for Application*, DOT HS 812 014, at: http://www.nhtsa.gov/staticfiles/rulemaking/pdf/V2V/Readiness-of-V2V-Technology-for-Application-812014.pdf (Last accessed June 2015)
[93] *National Connected Vehicle Field Infrastructure Footprint Analysis - Final Report (FHWA-JPO-14-125)*, AASHTO/USDOT; June 27, 2014 at http://stsmo.transportation.org/Documents/AASHTO%20Final%20Report%20_v1.1.pdf (last accessed June 2015).
[94] *Ibid.*

These milestones will be reassessed as research is completed and as State and local agencies develop deployment plans. For instance, the 2040 milestone for traffic signal instrumentation may be adjusted once State and local planning analyses are conducted,[95] but assumes an approximately 20-year signal equipment replacement cycle and phase-in of DSRC equipment as older equipment is retired. The 2040 milestone for instrumentation of additional roadway locations assumes initial targeting of existing ITS infrastructure sites—locations where Connected Vehicle applications may augment or replace existing traveler information dissemination and/or data collection equipment.

These percentages also assume that USDOT pursues (a) rulemaking for deployment of 5.9 GHz DSRC on-board equipment in light vehicles in support of V2V safety applications, and (b) makes a comparable decision on heavy vehicles in subsequent years. The roadmap is then based on a range of vehicle penetration based on market sales beginning in 2020. The roadmap also incorporates the AASHTO Footprint DSRC RSU deployment projections.

The roadmap is predicated primarily on a market-driven scenario wherein applications are provided by industry for consumer adoption and public agency implementation. Along with selective infrastructure investments, public agencies will establish the necessary policy and regulatory foundations to enable public-private partnerships that result in deployment and operations that prioritize public safety and transportation management needs, but allow for innovative market opportunities in the areas of mobility and environmental services.

USDOT anticipates a Connected Vehicle infrastructure build-out process that is based on the national experience deploying earlier ITS applications, which was facilitated by:

- Expansion of existing test sites and early deployments;

- Leveraging the planning for end-of-life retirement of traffic signal controllers to implement next generation ITS-ready equipment;

- Instrumenting existing infrastructure locations where new applications can improve upon existing data collection or information dissemination approaches at reduced cost by capitalizing on existing power and communications capability;

- Using the Highway Safety Manual and Highway Performance Management System to identify traffic safety and congestion locations where safety and congestion reduction benefits are likely to have the greatest impact; and

- Using major national and international freight corridors as areas for early deployment of Connected Vehicle freight applications.

---

[95] These analyses may include state, regional and/or corridor level Connected Vehicle deployment plans and warrant analyses at specific locations.

Given the preceding vehicle and infrastructure milestones and the historic experience with ITS deployment, a candidate path for implementation over the next 15 to 20 years can be illustrated as a series of distinct but overlapping advancements (automated vehicles are expected to proliferate over this same period).

Advancement 1 (2015-2025)

- The roadmap assumes that embedded cellular communications become increasingly common in vehicles and DSRC-equipped light vehicles appear on the streets at the rate of many millions per year starting in model year 2020. V2V safety benefits are possible immediately and increase continually as new DSRC-capable vehicles enter the national fleet.

- Early adopter states and regions—those currently participating in Connected Vehicle testing or deployment—expand existing and launch new deployments that can provide benefits without high concentrations of equipped vehicles.

- Early adopters take advantage of opportunities to replace retired traditional equipment with CV-capable units and to instrument major congested urban corridors and significant rural/intercity safety spot locations, leveraging existing power and communications infrastructure.

Advancement 2 (2020-2035)

- Embedded cellular communications are in most vehicles; and the percentage of the DSRC-equipped light vehicle fleet increases to 80% or more from 2030 to 2035, providing expanded, larger-scale V2V and V2I safety benefits and V2I benefits.

- Early adopters conduct "in-fill" implementation and expand to safety spot locations that lack existing infrastructure while other states and regions—those less active in past and current CV testing and deployment—will initiate implementation on major congested urban corridors and at significant rural/intercity safety spot locations.

- Major Interstate, freight-intensive corridors begin to fill in, linking previously instrumented large urban areas and rural/intercity spot locations.

Advancement 3 (2030-2040)

- The percentage of the DSRC-equipped light vehicle fleet increases to 90% or more and most vehicles also have embedded cellular communications capability.

- Other states and regions begin filling in their urban areas (more corridors; greater infrastructure density).

- Additional Interstate corridors are instrumented, linking major metropolitan areas and providing greater coverage of spot locations in other rural/intercity areas.

Figure 4-1, on the next page, illustrates these advancements on a timeline.

Chapter 4 Recommended Path for DSRC Implementation

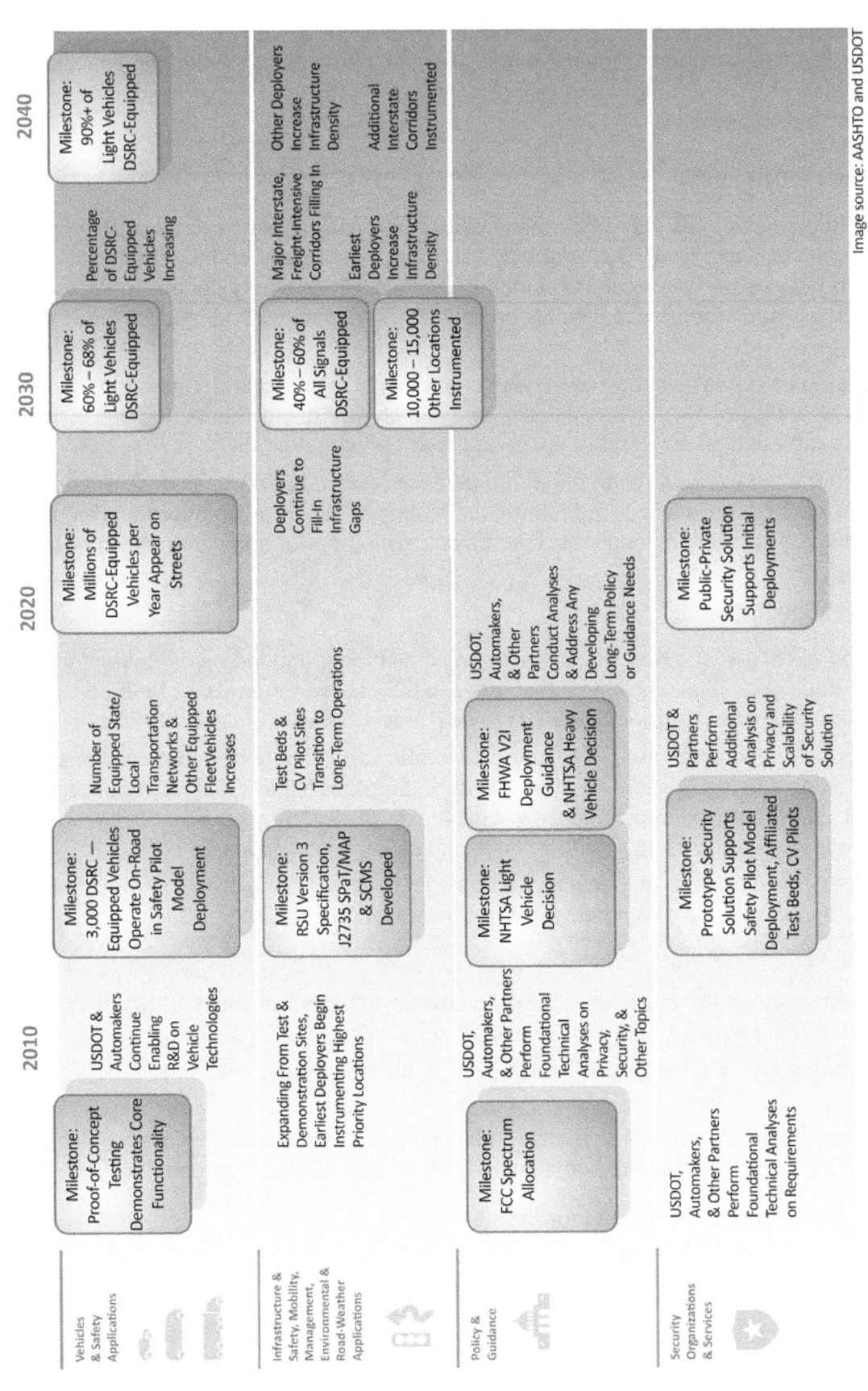

Figure 4-1. Candidate Paths to Deployment

4.II. Enabling Implementation

To implement this candidate roadmap requires other needed guidance and actions to enable implementation of DSRC and other components of the Connected Vehicle environment. These products include those pertaining to licensing, system architecture and standards, certification and policies. Each of these is discussed below.

4.II.A. Funding

Funding the deployment, operation, and maintenance of DSRC roadside hardware and software may add to the financial challenges of transportations agencies, but will also provide another tool to improve the safety and operations of the transportation system. To help agencies that elect to deploy V2I, FHWA will be releasing V2I Deployment Guidance and Products starting in the summer of 2015 that will include material on Federal-Aid eligibility for V2I activities. In short, this document will clarify the eligibility of certain Federal-Aid programs for V2I deployment, to support and encourage greater flexibility in the use of Federal funding for V2I implementation. The guidance also notes opportunities during the life cycle of legacy transportation management systems to replace (or upgrade) equipment with new V2I-capable equipment and software applications. This use of routine upgrades to capitalize on new features is envisioned as a low-cost pathway to full implementation.

Although many of the busiest intersections—which would be priority candidates for V2I infrastructure—are at the municipal level, Federal-Aid funds that flow to States also flow, through the planning process, to local level agencies. As local agencies plan for updating/replacing current systems, Connected Vehicle technologies will provide them with additional capabilities.

Based on the successful results of the Connected Vehicle research program, and the recent decision by NHTSA to pursue DSRC-enabled safety technology for light vehicles, a robust Connected Vehicle pilot program is envisioned as a mechanism to spur the implementation of Connected Vehicle technology. These pilot deployments will serve as initial implementations of Connected Vehicle technology in real world settings, with the aim of delivering near-term safety, mobility, and environmental benefits to the public. Pilot deployments offer an opportunity for stakeholders and partners to develop operational systems that exist well beyond the life of the program.[96] Upcoming awards to two waves of CV Pilot sites—in 2015 and in 2017—are anticipated to provide funding for initial DSRC infrastructure demonstrations.

---

[96] For additional information on pilot deployments, see: http://www.its.dot.gov/pilots/index.htm. (Last accessed June 2015)

4.II.B. Licensing

To support successful deployment of Connected Vehicle environments, implementing agencies will need guidance and supporting reference materials for licensing and siting of DSRC roadside equipment operating in ITS Radio Service, as defined in FCC rules, parts 90 and 95. At this time, the FHWA is developing guidance to assist State DOTs with decisions regarding deploying and implementing Connected Vehicle environments. Other modal administrations are expected to provide a basis to their constituents as well.

To date, a range of materials describes the general process and requirements for obtaining DSRC licenses from the FCC as well as the terms and conditions for maintaining licenses over the equipment's lifecycle. There also exists a host of information and lessons learned from existing test sites (i.e., Safety Pilot Model Deployment, Southeast Michigan test bed, and some of the State-based test beds such as Virginia's deployment along I-66).

In 2015, FHWA is leading the synthesis of existing materials into a draft guide on DSRC licensing through which FHWA will provide to implementers. The guide is expected to be focused from a deployer/implementer's perspective, which includes:

- Administration—how to obtain a license, how to address interference, and coordination between adjacent jurisdictions (important topics and requirements to address);
- Management—how to maintain a license, how to detect interference, and the necessary equipment or services for an agency to have available to detect interference;
- Field—significant parameters and guidance on design, siting, placement, and choosing a location; ensuring optimal performance of the radio; testing and certification of equipment; and
- Rule/Regulation—the roles and responsibilities of an agency with regard to the FCC rule/regulation and the identification of the limitations or places where the regulation might evolve and impact an agency in the future.

4.II.C. Role of the National ITS Architecture

The purpose of the National ITS Architecture is to enable State and local agencies to meet their local needs in planning for ITS deployments while ensuring nationwide interoperability.[97] The

---

[97] On January 8, 2001, USDOT issued a final rule to implement section 5206(e) of the Transportation Equity Act for the 21st Century (TEA–21), enacted on June 9, 1998, which required ITS projects funded through the highway trust fund to conform to the National ITS Architecture and applicable standards. Because it is highly unlikely that the entire National ITS Architecture would be fully implemented by any single metropolitan area or State, this rule requires that the National ITS Architecture be used to develop a local implementation of the National ITS Architecture, which is

National ITS Architecture is offered as guidance—standards and requirements are prescribed only to the level needed to meet interoperability.[98]

One of the key benefits in developing an ITS architecture is the ability to maximize the use of existing funding; a system developed for traffic management has the ability to share data (with appropriate agreements) with other transportation agencies thus eliminating redundancy and cutting costs. USDOT provides tools to support regional and statewide planning (for instance, the Turbo Architecture tool[99]) and offers technical assistance, training, and deployment support to agencies.

In 2012, recognizing the rapid development and complex nature of Connected Vehicle environments (particularly due to the cooperative nature of the applications), USDOT initiated the development of a reference architecture at a greater level of detail to support deployment. This CVRIA is informed by ongoing US and global research, and will evolve as results of this research are delivered. Once sufficiently mature, the CVRIA will become a part of the National ITS Architecture. A stable CVRIA is able to offer implementers at least one, if not more, approaches for how to implement Connected Vehicle applications and technologies. Development of the CVRIA has included and will continue to include robust stakeholder engagement and feedback. As new applications are conceived and as the interfaces are standardized and systems are deployed, the lessons learned from those projects will be fed back into the CVRIA to ensure that the Connected Vehicle environment is able to evolve and grow gracefully. To assist the stakeholder community in developing Connected Vehicle architectures, USDOT is making software tools available (the SET-IT tool[100] has recently been released in an alpha version) and will continue to support deployment activities, including new training that is under development.

4.II.D. Role of ITS Standards

USDOT's ITS Standards program, informed by the CVRIA, has initiated analysis of the CVRIA interfaces to identify candidates for standardization. Some interfaces may already be

---

referred to as a "regional ITS architecture." Conformance with the National ITS Architecture is defined under this rule as development of a regional ITS architecture and the subsequent adherence of ITS projects to the regional ITS architecture. The regional ITS architecture is based on the National ITS Architecture and consist of several parts including the system functional requirements and information exchanges with planned and existing systems and subsystems and identification of applicable standards, and would be tailored to address the local situation and ITS investment needs. http://www.ops.fhwa.dot.gov/publications/regitsarchguide/raguide.pdf. (Last accessed June 2015)

[98] Note the legislative requirement in SAFETEA-LU, SEC. 5307. NATIONAL ARCHITECTURE AND STANDARDS, (2) INTEROPERABILITY AND EFFICIENCY. (at: http://www.gpo.gov/fdsys/pkg/PLAW-109publ59/pdf/PLAW-109publ59.pdf, p. 668. (Last accessed June 2015)

[99] The tool is located at: http://www.iteris.com/itsarch/html/turbo/turbomain.htm. (Last accessed June 2015)

[100] http://www.its.dot.gov/arch/set_it.htm. (Last accessed June 2015)

standardized, others may need some modification to accommodate a new interface or concept, and some interfaces may require new standards or no standards at all.

A Standards Plan is under development and, with stakeholder input and cooperation with the Standards Development Organizations (SDOs),[101] a set of priorities and timelines for addressing standardization will be offered to USDOT. This input will assist USDOT in determining how to most effectively support standards needs in the future, including identifying opportunities for internationally harmonized standardization.

The essential Connected Vehicle standards for DSRC safety technologies and applications have been discussed previously in this report. Appendix D: History of DSRC: Policy and Technical describes the standards and offers information about their status and availability.

With regard to the significance of standards, the AASHTO Footprint[102] analysis makes the following observations:

- *Technical standards specify DSRC RSU form/fit/function and OBE function. Developers and deployers will eventually need stable technology standards for DSRC equipment to justify their investment of resources in application development and deployment. Lack of such standards could result in a reduced pace of deployment, or in variations in technical specifications among vehicle manufacturers and agencies.*

- *Technical standards specify interfaces and messages between vehicles and infrastructure. A minimum set of messaging standards will need to be implemented to support the Connected Vehicle applications. The vehicle will provide basic information about its location and speed; its sensed vehicle and road environmental conditions; and relevant requests for interactions with the infrastructure (e.g., traffic signals). The infrastructure will provide traveler information; maps and roadway geometries; and signal phase and timing. The infrastructure may also provide requests for data sensed by vehicles; roadside traffic and weather alerts; reference positioning corrections, and reference time corrections. Lack of such standards could result in a reduced pace of deployment, or in variations in technical specifications among vehicle manufacturers and agencies.*

- *Technical standards specify interfaces and messages between the roadside infrastructure and network information services. It would certainly be possible to build a*

---

[101] SAFETEA-LU, SEC. 5307. NATIONAL ARCHITECTURE AND STANDARDS, (3) USE OF STANDARDS DEVELOPMENT ORGANIZATIONS (at: http://www.gpo.gov/fdsys/pkg/PLAW-109publ59/pdf/PLAW-109publ59.pdf, p. 668. (Last accessed June 2015)

[102] Materials can be found at: http://stsmo.transportation.org/Documents/Executive%20Briefing.pdf and at http://stsmo.transportation.org/Documents/Task%206a%20AASHTO_CV_Footprint_Deployment_Scenarios_v2.pdf. (Last accessed June 2015).

set of closed systems where the roadside infrastructure was linked only to specific network nodes and services, but it would severely limit the opportunities for third-party application development and sharing of the roadside infrastructure among multiple applications. The alternative is, as assumed here, to establish standardized interfaces.

Stability of Standards

USDOT has been supporting an aggressive program to finalize stable standards in support of NHTSA V2V rulemaking activity. The key standards required to support the V2V over-the air-interface include IEEE 802.11p/1609 and SAE J2735 (now being split into J2735 and J2945.) These standards were developed specifically to support V2V and V2I wireless interfaces. They establish a wireless link for V2V and V2I communications (IEEE 802.11p), establish protocols for information exchange across the wireless link (IEEE 1609.x), and define message content for communicating specific information to and from equipment and devices via DSRC (SAE J2735 and SAE 2945.x) or other communications media.

These standards are stable in the areas most critical to ensuring interoperability of Wireless Access for Vehicular Environments (WAVE) communications, and have been demonstrated during the Safety Pilot model deployment to be well-suited to support V2V deployment. An aggressive effort is currently underway to refine the content of these standards to support large scale deployment, with stable content expected to be available by the end of 2015; publication to follow shortly thereafter.

USDOT expects to continue to detail standards needs for full-scale deployment, via the CVRIA program. This analysis can be performed in cooperation with international partners to leverage knowledge and budgets to:

- Identify a comprehensive set of standards needed for a Connected Vehicle environment;
- Identify gaps or identify where existing standards can be adopted or modified to meet needs; and
- Prioritize support to complete development of needed additional standards.

While stability of industry consensus standards are desirable in advance of a promulgation of an FMVSS, such stability is not absolutely required. If needed as a fallback option, the FMVSS can define in detail the minimum requirements and performance needed to ensure interoperability. Part of ensuring interoperability is making sure that DSRC exchanges information the same way every time and uses standardized messages. Each of the standards discussed in this section facilitates some part of DSRC operation.

Through the CVRIA program, along with planned international cooperation and work within USDOT's ITS Architecture and Standards Programs, USDOT expects to continue to document needs for standards across the envisioned full-scale deployment Connected Vehicle environment and prioritize USDOT support to complete development of needed additional standards. The content of standards continues to be developed via consensus processes in cooperation with SDOs and in accordance with legislative direction and good practice. Standards development will also continue to be informed by a diverse group of interested industry, academic and governmental stakeholders as well as via lessons learned from research, testing, and pilot deployment activity sponsored by USDOT and other organizations.

USDOT recognizes that maintaining forward and backward compatibility of standards across a decades-long envisioned deployment lifecycle is critical to obtain the benefits of fleet-wide interoperability. Updates that are currently taking place will intentionally break backward compatibility in a coordinated way in order to take full advantage of lessons learned from research and model deployments. Once the current revision cycle is complete (USDOT anticipates stable content by the end of 2015) and in advance of deployment, the USDOT's intent is to maintain full backward and forward compatibility, absent exceptional circumstances, through subsequent revision cycles.

4.II.E. Certification

Connected Vehicle Certification processes are critical for ensuring device performance, interoperability, and safety. With certification approval for technologies that cover communications, security, and applications, the envisioned Connected Vehicle Certification solution will be able to verify that the technologies participating in Connected Vehicle environments do so in a trusted, secure, and interoperable manner.

In the 2003 Report and Order, the FCC:

> ...requires that DSRC equipment operating in the 5.9 GHz band be certified in accordance with the procedures in Parts 2, 90, and 95 of the Rules, because these devices will be widely deployed and non-compliance with requirements could cause serious interference problems. Consequently, [FCC] requires all transponders, transmitters, and transceivers, whether associated with RSUs or OBUs used in the DSRCS to be certified...test procedures to demonstrate compliance with the standard shall also be left to industry to determine how to best achieve. To ensure compliance, applicants will be required to supply a statement that the equipment was tested and complies with the ASTM-DSRC Standard, as a prerequisite for certification. [103]

"Certification" implies a technical process as well as policies and procedures. To identify the parameters for policy, the technical processes and the elements (systems, devices, etc.) need to be analyzed to understand where certification is most relevant; the options for how certification processes might work; and whether there are tools or other resources needed to perform certification that are not in existence today. Concurrently, from the policy side, analysis is required to determine the elements of a comprehensive and targeted certification policy; to determine the needs and risks that certification resolves; to identify the legal and institutional

---

[103] p.26 at: http://hraunfoss.fcc.gov/edocs_public/attachmatch/FCC-03-324A1.pdf. (Last accessed June 2015)

challenges; and to determine the resources required for implementation of the policy. Importantly, certification policy decisions will have significant and long lasting effects on the Connected Vehicle environment as well as business models and opportunities.

A key objective in developing a certification policy is to ensure that certification is applied where it is most needed, and to also ensure that the policies do not burden the process.

The USDOT has initiated Certification pilot tests that will involve public-sector and private-sector entities. Three independent testing organizations[1] are participating in the pilots; they are developing certification processes for key information flows within the connected vehicle reference implementation architecture (CVRIA). Pilot test site teams will define the capabilities needed in their installations, using guidelines given by the USDOT. Hardware and software developers will then create devices and applications that embody those desired capabilities. Certification teams will independently evaluate the hardware and software, or the developers will self-certify.

4.II.E.1   Proposed Certification Paths

Certification activities related to V2X devices are proceeding at USDOT along two paths. First, NHTSA is developing vehicle-level test procedures that will be used to verify compliance with functional, operational and performance requirements to be detailed in the NPRM. The focus of such vehicle-level testing procedures will be to ensure interoperability among DSRC devices as installed in the vehicle, and that they meet the performance levels needed to support the safety applications.

Second, USDOT is also helping to foster a market-based certification ecosystem that can be leveraged by various suppliers within the Connected Vehicle supply chain to ensure interoperability and performance levels of stand-long components and sub-systems. In August 2013, the ITS Joint Program Office issued a Request for Information (RFI) that proposed a four-stage process for granting device and system certification. Stakeholders provided feedback on the process and, more importantly, about the existing state of the industry and the perceived challenges and opportunities for establishing these new certification stages/processes.

A June 2014 Request for Applications (RFA) for *Connected Vehicle - Next Stage Certification Environment* has resulted in three partnerships with industry laboratories to advance their practices to be able to sustainably provide certification services for Connected Vehicle technologies and applications. To arrive at this result, further analysis is needed that will lead to:

---

[1] DanLaw, OmniAir Certification Services, and 7Layers

- Development and documentation of policies, plans, procedures, and tools that will be needed to conduct qualification and certification testing for various devices and applications used in large-scale deployment trials to be conducted over the next few years;

- Results from tests conducted on various devices and the applications (testing will include radio and message level interface testing, as well as some system performance testing (e.g., radio range and positioning accuracy)); and

- Conclusions and actions for establishing a Connected Vehicle Certification Testing program that can be financially and institutionally self-sustaining.

By issuing awards as grants, USDOT has the ability to work in a research capacity with industry entities that are building specialized test equipment and changing organizational processes. For those entities, early financial sustainability will come from support with testing in association with the Connected Vehicle Affiliated Test Beds and planned Connected Vehicle Pilots. For USDOT, lessons learned will inform policies and Guidance, identify which new standards and test procedures will be needed by industry, and result in tools and reference materials.

This work began in February of 2015. Working under Cooperative Agreements, the three facilities will conduct qualification and certification testing for various connected devices and applications. In the process of developing certification procedures, USDOT is using Communication Views of the reference architecture as a guide for creating certification modules. The teams are following a well-established layered approach to constructing the communications (e.g., the OSI 7 Layer model). Certification entities will be able to break up the objects that make up the system and create certification processes for each of the layers starting with the various physical media and building up to the application layer at the top.[104]

---

[104] More detail on the Certification research can be found at:
http://www.its.dot.gov/connected_vehicle/connected_vehicle_cert.htm. (Last accessed June 2015)

Chapter 5 Conclusion

DSRC is the only wireless technology that provides trusted and secure, low latency, wireless data communications that meet the unique needs of high-speed data exchange among moving vehicles and with roadway infrastructure devices without compromising personal privacy or facilitating the tracking of traveler whereabouts. These attributes are essential to safety-critical crash avoidance technologies and applications. These technologies are ready for deployment.

While analysis continues to conclude that DSRC is, to date, the only viable option for safety-critical applications, these same efforts identify opportunities to use non-DSRC communications in support of applications that do not require the low latency, high reliability and availability of the DSRC band, or privacy protections.

In summary, this report concludes the following:

- **DSRC remains a critical requirement for enabling safety-critical V2V and V2I applications.** Vital characteristics are:
  - Rapid transmission speed, low latency, stability, and dedicated availability make DSRC the optimal communication media currently available for safety-critical applications, particularly those focused on crash avoidance.
  - The highest levels of privacy are achieved through system designs that do not use or store any personal information.

- **Operations that use DSRC—test beds, operational sites, and emerging pilot sites—are demonstrating how the spectrum is used**; specifically finding that:
  - The FCC's service rules and spectrum band plan are viable. Additional details regarding licensing and conflict management of jurisdictional installations are undergoing analysis to explore options; and are expected to be presented to the FCC within the 2016-2017 timeframe.
  - Non-DSRC communications technologies are viable for mobility, environmental, and other non-safety critical applications that do not require the specific characteristics of DSRC. However, since these other technologies do not afford users the same level of privacy that DSRC technologies do, users may not enjoy the same level of privacy protection and, therefore, may be required to "opt-in" to applications using non-DSRC technologies.

- **DSRC is ready for wider-scale implementation**:
  - USDOT has synthesized stakeholder deployment plans into a roadmap to illustrate a candidate path for implementation
  - Vendors and standards working groups have taken results of the Safety Pilot Model Deployment tests and are improving and refining their products to be production-ready and market available.

- o USDOT is preparing the policy foundation:
  - ▪ NHTSA will issue a NPRM with regard to DSRC-based technologies for light vehicles in early 2016;
  - ▪ FHWA is issuing Guidance for State and local agencies along with tools and reference guidelines; and
  - ▪ The ITS JPO is working on a comprehensive Connected Vehicle standards plan; is preparing for integration of the Connected Vehicle Reference Implementation Architecture into the National ITS Architecture; has produced an architecture tool for State and local implementers; and is continuing to pursue opportunities to harmonize its architecture and standards on an international level.
  - o An initial security credential management system will be available in late 2016 to support early deployments.
  - o USDOT is also launching new Connected Vehicle Pilot Deployments that are meant to serve as a foundation for state and local deployments.
- **With regard to recently emerging "unknowns" regarding coexistence with unlicensed devices**—completion of analysis and testing in 2016, and simulation modeling in 2016-2017 will provide the details necessary to further inform the FCC's exploration of sharing technologies on whether:
  - o Co-existence of Connected Vehicle technologies with unlicensed devices is proven feasible.
  - o Proposed spectrum sharing plans or technologies can prove the capability to avoid harmful interference with crash-avoidance safety systems. Upcoming tests will provide necessary insights to inform policy.

Throughout the ongoing research, USDOT and its partners have evaluated how the allocated DSRC spectrum will be applied. As of mid-2015, approximately 20 implementing agencies and some academic and private sector organizations have licenses that allow for use of DSRC for experimental installations. These licensed sites will form the basis for initial operations in those areas. Descriptions of these sites are provided in this report in Appendix C. They show the interest in advancing to this next generation of ITS by agencies in: Arizona, California (multiple sites including a transit operations site in San Francisco), Florida, Michigan (multiple sites), New York, Virginia (multiple sites), and Washington; and across the border in Canada (Alberta and British Columbia), where there are ongoing discussions about architecture harmonization.

As evidenced by this report, DSRC is ready for use and DSRC-based technologies and applications offer a path to a safer and more efficient surface transportation system for America.

References

Ahmed-Zaid, F., Bai, F., Bai, S., Basnayake, C., Bellur, B., Brovold, S., et al. (2011). *Vehicle Safety Communications – Applications (VSC-A) Final Report.* CAMP and NHTSA, USDOT.

American Association of State Highway and Transportation Officials (AASHTO). (2014). *National Connected Vehicle Field Infrastructure Footprint Analysis - Final Report* (FHWA-JPO-14-125), AASHTO and FHWA, USDOT.

Andrews, S., Avercamp, J., Chuang, P., Colins, J., Frank, L., Freckleton, D., et al. (2013). *Development of DSRC Data Communication System Performance Measures.* U.S. Department of Transportation, Research and Innovative Technology Administration.

FCC Memorandum Opinion and Order, FCC 06-110 in the Matter of WT Docket No. 01-90, ET Docket No. 98-95, and RM-9096 (Federal Communications Commission July 26, 2006).

FCC Report and Order, FCC 99-305 in the Matter of ET Docket 98-95 and RM-9096 (Federal Communications Commission October 22, 1999).

FCC Report and Order, FCC 03-324 in the Matter of WT Docket No. 01-90, ET Docket No. 98-95, and RM-9096 (Federal Communications Commission February 10, 2004).

Fehr, W. (2014). ITS Program Manager for Systems Engineering.

ITS Joint Program Office. (Progress Update 2012). *Transforming through Connectivity: ITS Strategic Research Plan, 2010-2014.* U.S. Department of Transportation.

ITS Strategic Research Plan, 2010-2014. (n.d.). Retrieved from http://www.its.dot.gov/strategic_plan2010_2014

Misener, J., Andrews, S., Cannistra, P., & Garcia, D. (2012). *Communications Data Delivery System Analysis Public Workshop Read-Ahead Document.* U.S. Department of Transportation, Research and Innovative Technology Administration, Intelligent Transportation System Joint Program Office.

Najm, W. G., Koopmann, J., Smith, J. D., & Brewer, J. (2010). *Frequency of Target Crashes for IntelliDrive Safety Systems.* Cambridge, MA: U.S. Department of Transportation, Research and Innovative Technology Administration, John A. Volpe National Transportation Systems Center.

National Highway Traffic Safety Administration (NHTSA). (2014, February 3). *U.S. Department of Transportation Announces Decision to Move Forward with Vehicle-to-Vehicle Communication Technology for Light Vehicles*. Retrieved June 9, 2014, from www.nhtsa.gov: http://www.nhtsa.gov/About+NHTSA/Press+Releases/2014/USDOT+to+Move+Forward+with+Vehicle-to-Vehicle+Communication+Technology+for+Light+Vehicles.

NHTSA. (2000-2014). *Office of Crash Avoidance Research Technical Publications*. Retrieved January 30, 2014, from National Highway Traffic Safety Administration: http://www.nhtsa.gov/Research/Crash+Avoidance/ci.Office+of+Crash+Avoidance+Research+Technical+Publications.print.

NHTSA. (2010). *Frequency of Target Crashes for IntelliDrive Safety Systems.* U.S. Department of Transportation.

National Telecommunications and Information Administration (NTIA). (2012) *Case Study: Investigation of Interference into 5 GHz Weather Radars from Unlicensed National Information Infrastructure Devices, Part III.*Carroll, John E., Geoffrey A. Sanders, Frank H. Sanders, and Robert L. Sole. NTIA Report TR-12-486. www.its.bldrdoc.gov/publications/download/12-486.pdf.

SAE J2735-2009. (Rev 9.0, 2011; Rev. 11.1, 2012). System Requirement Specification for the 5.9 Ghz DSRC Vehicle Awareness Device Specification. *Model Deployment Device DSRC BSM Communication Minimum performance Requirements.*

Sen, B., Campbell, B. N., Smith, J. D., & Najm, W. (2002). *Analysis of Light Vehicle Crashes and Pre-Crash Scenarios Based on the 2000 General Estimates System.* Cambridge, MA: U.S. Department of Transportation Research and Innovation Administration John A. Volpe National Transportation Systems Center.

U.S. Department of Transportation Research and Innovative Technology Administration. (2014, March 18). *Principles for a Connected Vehicle Environment.* Retrieved from Intelligent Transportation Systems Joint Program Office: http://www.its.dot.gov/connected_vehicle/principles_connectedvehicle_environment.htm.

U.S. Government Accountability Office (GAO). (2013). *Intelligent Transportation Systems: Vehicle-to-Vehicle Technologies Expected to Offer Safety Benefits, but a Variety of Deployment Challenges Exist.*

Appendix A. Glossary of Acronyms

AASHTO	American Association of State Highway and Transportation Officials
ACTIVE	Alberta Cooperative Transportation Infrastructure and Vehicular Environment
ADOT	Arizona Department of Transportation
ANPRM	Advanced Notice of Proposed Rulemaking
ARINC	Aeronautical Radio, Incorporated (**ARINC**), established in 1929, now part of Rockwell Collins
ASD	Aftermarket Safety Device
ASTM	ASTM International, formerly known as the American Society for Testing and Materials
ATTRI	Accessible Transportation Technologies Research Initiative
AURORA	Automotive Test Bed for Reconfigurable and Optimized Radio Access
AVL	Automated Vehicle Location
BR/EDR	Bluetooth Basic Data Rate/Enhanced Data Rate (BR/EDR)
BSM	Basic Safety Message
BSW/LCW	Blind Spot/Lane Change Warning
CA	Certificate Authority
CACC	Cooperative Adaptive Cruise Control
Caltrans	California Department of Transportation
CAM	Cooperative Awareness Message
CAMP	Crash Avoidance Metrics Partnership
CAN	Controller Area Network
CCH	Control Channel
CDDS	Communications Data Delivery System
CICAS	Cooperative Intersection Collision Avoidance Systems
CME	Credential Management Entity
Cor.	Corrigendum, a correction of a published text
CRL	Certificate Revocation List
CSW	Curve Speed Warning
CV	Connected Vehicle

CVI-UTC	Connected Vehicle/Infrastructure University Transportation Center
CVII	Commercial Vehicle Infrastructure Integration
CVRIA	Connected Vehicle Reference Implementation Architecture
DENM	Decentralized Environmental Notification Message
DG CONNECT	Directorate General for Communications Networks, Content & Technology (EC)
DMS	Dynamic Message Signs
DNPW	Do Not Pass Warning
DoD	Department of Defense
DoS	Denial of Service Attack
DOT	Department of Transportation
D-RIDE	Dynamic Ridesharing
DSL	Digital Subscriber Line
DSRC	Dedicated Short-Range Communications (5.9 GHz)
DVI	Driver-Vehicle Interface
EC	European Commission
ECU	Engine Control Unit
EEBL	Emergency Electronic Brake Lights
ETSI	European Telecommunications Standards Institute
EU	European Union
EVAC	Emergency Communications and Evacuation
FAA	Federal Aviation Administration
FCC	Federal Communications Commission
FCW	Forward Collision Warning
FHWA	Federal Highway Administration
FMCSA	Federal Motor Carrier Safety Administration
FRATIS	Freight Advanced Traveler Information Systems
FMVSS	Federal Motor Vehicle Safety Standard
FSS	Fixed Satellite Services
FTA	Federal Transit Administration
GAO	Government Accountability Office
GHz	Gigahertz
GID	Geometric Intersection Design

GPRS	General Packet Radio Service
GPS	Global Positioning System
HD	High Definition
HOT	High Occupancy Toll
HTG	Harmonization Task Group
HV	Heavy Vehicle
I2V	Infrastructure-to-vehicle
ICT	Information and communication technologies
IEEE	Institute for Electrical and Electronics Engineers
IMA	Intersection Movement Assist
INC-ZONE	Incident Scene for Work Zone Alerts for Drivers and Workers
IP	Internet Protocol
I-SIG	Intelligent Traffic Signal System
ISM	Industrial, Scientific and Medical
ITS	Intelligent Transportation Systems
JPO	Joint Program Office
LA	Linkage Authority
LAN	Local Area Network
LLC	Logic Link Control
LOP	Location Obfuscation Proxy
LTA	Left Turn Assist
LTE	Long Term Evolution
LTE-A	Long Term Evolution - Advanced
M&O	Memorandum and Order
MA	Misbehavior Authority
MAC	Media Access Control
MAC	Medium Access Control
MAIS	Maximum Abbreviated Injury Scale
MAP	MapData
MAP-21	Moving Ahead for Progress in the 21st Century Act
MAW	Motorist Advisories and Warnings
MBD	Misbehavior Detection

Mbps	Megabit per second
MCDOT	Maricopa County Department of Transportation
MDOT	Michigan Department of Transportation
MHz	Megahertz
MMITSS	Multi-modal Intelligent Traffic Signal System
MOU	Memorandum of Understanding
MUTCD	Manual on Uniform Traffic Control Devices
NEMA	National Electronics Manufacturers Association
NHTSA	National Highway Traffic Safety Administration
NPRM	Notice of Proposed Rulemaking
NRC	National Research Council
NTCIP	National Transportation Communications for ITS Protocol
NTIA	National Telecommunications and Information Administration
OBE	On-board Equipment
OBU	On-board Unit
OEM	Original Equipment Manufacturers
OST/R	Office of the Assistant Secretary for Research and Technology
OVW	Oversize Vehicle Warning
PATH	Partners of Advanced Transit and Highways
PED-SIG	Mobile Accessible Pedestrian Signal System
PHY	Physical Layer
PIA	Privacy Impact Assessment
PKI	Public Key Infrastructure
PII	Personally Identifiable Information
PLCC	Plastic Leaded Chip Carrier
PPP	Public Private Partnership
PPSG	Policy and Plans Steering Group
PREEMPT	Emergency Vehicle Preemption
PSID	Provider Service Identifier
Q-WARN	Queue Warning
R&O	Report and Order
RA	Registration Authority

RCC	Regulatory Cooperation Council
RCVW	Railroad Crossing Violation Warning
RESP-STG	Incident Scene Pre-Arrival Staging Guidance for Emergency Responders
RLVW	Red Light Violation Warning
RSE	Roadside Equipment
RSU	Roadside Units
RSZW	Reduced Speed Zone Warning
SafeTrip-21	Safe and Efficient Travel Through Innovation and Partnerships in the 21st Century
SCH	Service Channel
SCMS	Security Credential Management System
SDARS	Satellite Digital Audio Radio Service
SDO	Standards Development Organization
SES	Luxembourg-based satellite company founded in 1985
SET-IT	Systems Engineering Tool for Intelligent Transportation
SIA	Satellite Industry Association
SPaT	Signal Phase and Timing
SPD-HARM	Dynamic Speed Harmonization
SSGA	Stop Sign Gap Assist
SSVW	Stop Sign Violation Warning
SWIW	Spot Weather Information Warning
TCA	Transport Certification Australia
T-CONNECT	Connection Protection
TDWR	Terminal Doppler Weather Radar
T-DISP	Dynamic Transit Operations
TEA-21	Transportation Equity Act for the 21st Century
TMC	Traffic Management Center
TSP	Transit Signal Priority
UBC	University of British Columbia
UDP	User Datagram Protocol
ULS	Universal Licensing System
UMTRI	University of Michigan Transportation Research Institute

U-NII	Unlicensed-National Information Infrastructure
USDOT	United States Department of Transportation
UWB	Ultra-Wideband
V2I	Vehicle-to-infrastructure
V2V	Vehicle-to-vehicle
V2X	Vehicle-to-portable device
VAD	Vehicle Awareness Device
VDOT	Virginia Department of Transportation
VDT	Vehicle Data Translator
VII	Vehicle Infrastructure Integration
VIIC	Vehicle Infrastructure Integration Consortium
VIN	Vehicle Identification Number
VSC	Vehicle Safety Consortium
VSC-A	Vehicle Safety Communications - Applications
VTRW	Vehicle Turning Right in Front of Bus Warning
VTTI	Virginia Tech Transportation Institute
WAN	Wide Area Network
WAVE	Wireless Access for Vehicular Environments
Wi-Fi	Local area wireless technology
Wi-Fi FILS	Wi-Fi Fast Initial Link Setup
WiMAX	Worldwide Interoperability for Microwave Access
WxTINFO	Weather Response Traffic Information System

Appendix B Summary of Analyses of Communications Media Options

B.I. 2014: Comparison of Communications Features for Meeting V2V Crash-Avoidance Requirements[105]

There are many factors in determining if a communications technology is suitable for supporting vehicle to vehicle safety applications. Critical factors include minimum and maximum range; interference protection through dedicated, licensed frequencies; the ability to broadcast messages from one vehicle to all neighboring vehicles; the ability to accommodate large ad-hoc networks that change in real time; network capacity; and adequate performance at highway speeds. However, one factor that is crucial to crash avoidance applications is latency: the time it takes a source to generate a message and have it received and interpreted by the destination.

Communications technology used for safety applications will incorporate its "infrastructure" between the applications. This infrastructure is vastly different for different wireless technologies. For example, a classic cellular system will require communications from a vehicle to a nearby base station subscribed to by the source application, communications from the base station through a gateway to a backhaul network, and subsequent communications through the backhaul network to a gateway serving a base station subscribed to by the destination application. Cellular communications for applications such as V2V safety may also involve accessing a server, because trajectory information used for collision avoidance calculations should be restricted to vehicles within a short distance from each to avoid overwhelming vehicles with messages from other vehicles too far away to be of interest (DSRC achieves this automatically through its short range nature). Each component in the chain will contribute to the end-to-end latency.

Several technologies have either developed or proposed a "direct" device to device mode of communications, but they were not designed for constantly changing ad-hoc networks

---

[105] Extracted from a Noblis, Inc. white paper, dated July 10, 2014; provided as an update to a previous CAMP analysis referenced in the *Vehicle Safety Communications Project Task 3 Final Report Identify Intelligent Vehicle Safety Applications Enabled by DSRC*, CAMP report to NHTSA, March 2005.

consisting of numerous users traveling at highway speeds. In many cases they are not scalable nor do they have the transmission range to accommodate ITS applications.

The following table (Table B-1) summarizes some of the features of communications technologies that have been examined for use in ITS applications and how their capabilities compare with the needs for V2V safety applications. However, due to the many variations and choices that can be made by the companies that deploy these technologies, not all of the previously discussed factors are included. Dark blue on the table indicates that the need is met, gray that the need is not met, and light blue either that there is uncertainty in whether or not the technology could meet the need or that it is met in a less-than-optimal way, (e.g., wide-area communications creating a need for location-based filtering, as described above). Only DSRC meets all the needs of the V2V safety applications.

Appendix B. Summary of Analyses of Communications Media Options

Table B-1. Comparison of Two-way Communications Technologies with Selected V2V Requirements

Technologies:	Latency[1] (seconds)	Range[2] (meters)	Mobility Design Goal (Relative Speed)	Direct Device-to-Device Mode	Dedicated and Licensed Spectrum
V2V Safety Needs	0.1 s	300m	Highway speed	Y	Y
DSRC	0.0002 - 0.015s [3]	<1000 m	Highway speed	Y	Y
Wi-Fi	3 - 5 s	<300 m	Pedestrian speed	Y	N
Wi-Fi Direct	5 s	<200 m	Pedestrian speed	Y	N
Wi-Fi FILS draft standard	>0.1 s	<300 m	Pedestrian speed	N	N
LTE Cellular	0.079 - 0.1 s	Wide Area	Highway speed	N	Y
LTE-A Cellular	0.02 - 0.07 s	Wide Area	Highway speed	Proposed	Y
BlueTooth BR/EDR	0.1 s	<100 m	Stationary	Y	N
BlueTooth LE	0.006 s - 70 min	<50 m	Stationary	Y	N
Satellite	> 0.015 - 0.58 s [4]	Wide Area	Varies	N	Y

Table B-1 Notes:

1. In certain technologies, latency may include scanning, association and authentication, group discovery and formation, and transition from idle to connect state. Latency will increase with many factors, some of which include mobility, system loading, smaller channel bandwidths, and the configuration of infrastructure equipment (where appropriate). USDOT testing at Turner-Fairbank Highway Research Center showed a 1-2 second latency to establish a session on LTE. And, as described further on page 100, while the highest achievable future LTE data rate for moving users is 100 Mbps, in practice, because of user capacity limitations and interference, this is typically substantially lower.

2. Range depends on many factors including effective transmit power, system loading, interference from other devices and speed of travel by the user device.

3. The lower end of this range applies to one-way broadcast of V2V safety messages at short distances (50m). The higher end applies to Infrastructure to vehicle transmissions at approximately 800m.

4. These numbers represent a theoretical minimum round trip propagation time depending on satellite orbit (Low Earth Orbit to Geosynchronous), and only when the serving satellite is directly overhead. Actual latency will be significantly greater depending on many factors. Current Low Earth Orbit systems *might* have low enough latency, but low bandwidth (not discussed here).

Table B-2. Citations

Technology	Citations
DSRC	• *Wireless Communications Performance Based on IEEE 802.11p R2V Field Trials*, Jia-Chin Lin, Chi-Sheng Lin, Chih-Neng Liang, and Bo-Chiuan Chen, IEEE Communications Magazine, May 2012, Vol 50, No. 5. • *Vehicle Safety Communications Project Task 3 Final Report Identify Intelligent Vehicle Safety Applications Enabled by DSRC,* CAMP report to NHTSA, March 2005.
Wi-Fi	• *Vehicle Safety Communications Project Task 3 Final Report Identify Intelligent Vehicle Safety Applications Enabled by DSRC,* CAMP report to NHTSA, March 2005.
Wi-Fi Direct	• *Device to Device Communications with Wi-Fi Direct: Overview and Experimentation*, Daniel Camps-Mur, Andres Garcia-Saavedra and Pablo Serrano, IEEE Wireless Communications, Volume:20 , Issue: 3 June 2013
Wi-Fi Fast Initial Link Setup (FILS)	• *TGai Requirements Document,* Marc Emmelmann, May 12, 2011, http://www.ieee802.org/11/Reports/tgai_update.htm

Technology	Citations
DSRC	• *Wireless Communications Performance Based on IEEE 802.11p R2V Field Trials*, Jia-Chin Lin, Chi-Sheng Lin, Chih-Neng Liang, and Bo-Chiuan Chen, IEEE Communications Magazine, May 2012, Vol 50, No. 5. • *Vehicle Safety Communications Project Task 3 Final Report Identify Intelligent Vehicle Safety Applications Enabled by DSRC,* CAMP report to NHTSA, March 2005.
LTE	• http://www.fiercewireless.com/special-reports/3g4g-wireless-network-latency-comparing-verizon-att-sprint-and-t-mobile-feb (last accessed on June 24, 2014)
LTE-A	• *LTE CA: Carrier Aggregation Tutorial*, Ian Poole, http://www.radio-electronics.com/info/cellulartelecomms/lte-long-term-evolution/3gpp-4g-imt-lte-advanced-tutorial.php (last accessed on June 24, 2014) • Note: ITU goal for LTE-A: Latency: from Idle to Connected in less than 50ms and then shorter than 5ms one-way for individual packet transmission. Latency of 20ms after connection (active state) often quoted for round trip time if server is very close to the base station.
Bluetooth Basic Data Rate/Enhanced Data Rate (BR/EDR)	• *Bluetooth Technology-Advanced.* Brian Redding, Qualcomm Atheros presentation at Bluetooth Asia, Shanghai China, 2013.
Bluetooth Low Energy (LE)	• *Bluetooth Low Energy Version 4.0; Helping to Create the Internet of Things*, Julio Vellagas http://home.eng.iastate.edu/~gamari/CprE537_S13/project%20reports/Bluetooth%20LE.pdf (last accessed on June 24, 2014)
Satellite	• *Minimizing Latency in Satellite Networks*, Greg Berlocker, http://www.satellitetoday.com/telecom/2009/09/01/minimizing-latency-in-satellite-networks/ (last accessed on June 24, 2014)
V2V Safety Needs	• *Vehicle Safety Communications Project Task 3 Final Report Identify Intelligent Vehicle Safety Applications Enabled by DSRC,* CAMP report to NHTSA, March 2005.

B.II. 2012-2014: Independent Analyses on Communications Technologies

B.II.A. 2012-2013: For Security Credential Management by Booz Allen Hamilton

In addition to its low latency, DSRC provides the communications and network availability necessary for the management of security. This management includes the capability to:

- Provide the initial security credentials which identify a device as authenticated within the system (known as "boot-strapping" a device)
- Provide the initial batch of credentials which may last from 1-3 years
- Update and renew those credentials as needed
- Distribute a Certificate Revocation List (CRL), which is the largest and most frequent amount of data being transferred as part of security credentials management. The CRL provides instruction to devices to ignore the data emanating from the misbehaving actors within the system which are described in the list. The size of the CRL is an unknown as it will vary based on the number of attackers at any given time (known as the levels of misbehavior rates). Note that the more frequent the certificate update that is needed, the lower the rates of CRL transmission and the smaller the size of CRLs.

During the 2012-2013 timeframe, USDOT launched independent analyses on communications technologies to examine:

- Which technology or combinations of technologies are best suited as a medium for credential management
- Requirements for a nationwide communications data delivery system (CDDS) to support credential management
- Development of DSRC operational needs and performance requirements.

To perform these analyses, USDOT engaged Booz Allen Hamilton, who analyzed six over-the-air communication modes to determine their abilities to provide the needed communications that would operate within three Connected Vehicle environments: a V2V environment, a V2V/V2I environment, and a security credential management system. The six modes examined were:

- Cellular including 3G and 4G (or Long-Term Evolution; LTE)
- Wi-Fi
- DSRC
- Worldwide Interoperability for Microwave Access (WiMAX)
- Satellite Digital Audio Radio Service (SDARS)
- High Definition (HD) Radio

Three modes (WiMAX, SDARS, and HD Radio) were eliminated due to initial analysis that revealed they would not provide the needed support and network availability at the scale of the credential management system or communications with OBEs. The remaining three (Cellular, Wi-Fi, and DSRC) were evaluated in more detail to begin the process of understanding the business and institutional arrangements that may underpin delivery of certificate data.

Examination of options resulted in three scenarios that can provide viable data delivery:[106]

- Scenario 1: A Cellular-based system for security credentials management combined with a small percentage of DSRC for delivering V2V communications.
- Scenario 2: A Hybrid system for security credentials management that combines the use of cellular communications with opportunistic use of Wi-Fi and DSRC for security management and a small percentage of DSRC for V2V communications.
- Scenario 3: A DSRC-based system for security credentials management and for V2V communications.

Booz Allen Hamilton notes the following opportunities and limitations for credential management with each technology or combination of technologies:[107]

Cellular Technologies

Cellular systems are common throughout the nation and are continuing to expand. In particular, the advancement of LTE technology is helping to deliver larger amounts of data to cellular users more quickly. However, Booz Allen Hamilton's analysis stated that this is less effective when a user is moving, and that the data rate for LTE is often much lower than what is theoretically possible. Although LTE would be able to support the full download of credential revocation list due to the expansiveness of cellular networks, there are areas where cellular networks are not available. Even within coverage areas, there can be dead spots. Another issue that may arise is the fact that any LTE system may suffer from capacity issues in any area that has many LTE users. Though cellular could potentially be a viable option for coverage, the Booz Allen Hamilton research concluded that cost and security issues make it an unrealistic option for the CDDS.

Wi-Fi Technologies

Wi-Fi technology supports wireless connectivity and generally high data rates. The main drawback of Wi-Fi is its design for stationary terminals (i.e., vehicles that are parked). Though Wi-Fi offers higher data rates than do other options, it does not work nearly as well in a mobile environment. In addition, any vehicle that enters the Wi-Fi hotspot must give its MAC (media access control) address and obtain the MAC address of all other vehicles in the hotspot before it can send communications. Though it uses the same basic radio system as DSRC, DSRC eliminates the need for users to gather MAC

---

[106] Summarized from an internal USDOT technical memo to CAMP dated May 16, 2013.
[107] NHTSA: "Vehicle-to-Vehicle Communications: Readiness of V2V Technology for Application, DOT HS 812 014, pg. 264; and 266.at: http://www.nhtsa.gov/staticfiles/rulemaking/pdf/V2V/Readiness-of-V2V-Technology-for-Application-812014.pdf

addresses before communication. The costs and security risks associated with cellular also apply to Wi-Fi.

DSRC Technologies

DSRC cannot support a full CRL update (assuming a large fleet of vehicles with a misbehavior rate of 0.1%) if the vehicle passes a DSRC RSU at more than a few miles per hour. In order for a vehicle to receive a full CRL update, it must therefore pass by more than one DSRC RSU per day, and any update process would have to support incremental updates. The Booz Allen Hamilton results suggested that a typical system would require 40 seconds to complete a full CRL update, and a vehicle would only be in the footprint for 14 seconds in the absence of congestion.

However, the DSRC technology would be able to support incremental updates. And, the costs using the DSRC protocol are considerably lower than costs under the other two protocols. Furthermore, DSRC communication is believed to meet security requirements, thus it becomes a realistic choice for CDDS.

Satellite Radio Technologies

SDARS uses satellites to provide digital data broadcast service in a seamless manner across the Nation (approximately 98% of the US land mass) and may include 200 miles (322 km) of off-shore coverage.

SDARS is under reconsideration at this time for providing credentials, which can be done in a broadcast-mode. However, the Booz Allen Hamilton analysis suggested that SDARS may not be able to support the download of a full CRL because the download time would be longer than an average trip. If an incremental system is used, however, it could support updates. The costs and security risks associated with cellular also apply to satellite.

These four communications options were also analyzed by CAMP and the VIIC. The conclusion from the BAH analysis is that the DSRC scenario with a small network of DSRC RSUs ended up being the most economically viable as well as allowing for the most security. While Cellular and Hybrid show some merits, their costs and security concerns make them generally less attractive options.

B.II.B. 2013-2014: Communications Analysis for V2I Services by AASHTO[108]

The AASHTO analysis examined the characteristics of cellular systems and the typical cellular architectures that could be used to implement Connected Vehicle applications and systems. The analysis generally assumed that Connected Vehicle communications between mobile elements and field elements, particularly those supporting time sensitive V2I safety applications, would be carried out using DSRC technology; and that communications directly between Connected Vehicle mobile elements and center elements would be carried out using cellular/LTE.

A key difference in the AASHTO analysis of cellular communications is that the assumptions are based on needs of the Connected Vehicle systems and their interactions, rather than on the needs of specific applications. For the AASHTO analysis, a Connected Vehicle system is, in general, a means for:

- Communicating messages between mobile stations (e.g. vehicles)
- Communicating messages between field elements and mobile stations
- Communicating messages between center elements and mobile stations either directly or via the field elements[109]

Cellular systems are widely available and, driven by various consumer devices (smartphones, tablet computers, etc.). The cellular industry has been substantially expanding cellular capacity and coverage over the past 20 years.

The most recent advancement in cellular technology, known as LTE (long term evolution), is able to deliver very high data rates to fixed users; the highest achievable future LTE data rate for moving users is 100 Mbps. In practice, however, because of user capacity limitations and interference, this is typically substantially lower. LTE is backward compatible with previous generations, so in most cases new features that extend performance can be used without rendering earlier systems obsolete.

LTE is an all IP network. The cell areas are generally large, and each terminal is assigned an IP address when it joins the network. A variety of schemes have been developed to enable

---

[108] The following section is summarized from the draft *National Connected Vehicle Field Infrastructure Footprint Analysis: Deployment Concepts*, September 2013, pages 94-152, at http://stsmo.transportation.org/Documents/Task%206a%20AASHTO_CV_Footprint_Deployment_Scenarios_v2.pdf. (Last accessed June 2015).
http://stsmo.transportation.org/Documents/AASHTO%20Final%20Report%20_v1.1.pdf (last accessed June 2015).
[109] The Connected Vehicle environment includes mobile terminals, field terminals and center terminals. Mobile terminals are typically vehicles, while field terminals, when they are used, are typically radio terminals located along the roadway (typically called "roadside equipment", or RSE). Center facilities include traffic management centers and other road authority/agency back office facilities, and remote service providers.

terminals to maintain IP connectivity with remote servers as they move from cell site to cell site. As a result LTE may be suited to connecting mobile terminals to remote servers, but the network association time with each cell site still takes longer than is suitable for crash avoidance applications. And, contacting mobile terminals over the IP network is somewhat more complex, although mechanisms for this have been developed.

Unlike DSRC, LTE currently provides no provision for one mobile terminal to communicate directly with another nearby mobile terminal (or multiple mobile terminals) in a broadcast mode or with a local data source (e.g., a system that might be connected to an RSE to provide localized data). With LTE, all communications currently must go through the cellular system carrier's back haul network (a network that connects the cell site to the carrier's back office systems, and generally, to the Internet) and must include an IP address. An emerging addition is a system known as LTE-Direct. This system will allow communication directly between LTE terminal devices. It uses a concept known as "Proximate Discovery" that allows LTE terminals to announce the services they have to offer to other terminals in the local area. These announcements can then lead to one terminal providing information to other terminals in the area. The technology has not been widely used as yet, but it may someday be able to provide an LTE-based mechanism for V2V and V2I communications. It is early in the development of the technology; and many questions exist about whether carriers will find value in adopting these capabilities.

LTE uses a modulation approach that effectively creates multiple parallel radio links, so the overall rate of the combined set of links is higher than any single link. The approach, however, requires significant processing to adapt the system to the current radio propagation characteristics. Since these are dependent on the physical environment, the technique does not work as effectively when the user is moving.

For this reason, the highest achievable LTE data rate for moving users is only about 30% of the best stationary data rate. Still, LTE is a rapidly evolving technology.

An important difference between the Cellular/LTE system and DSRC is that in the current cellular system, mobile cellular terminals (e.g. handsets) establish a connection with a cell site (typically the closest, and/or the one with the strongest signal). This connection includes a subnet IP address so that the cell site can route IP packets to and from the mobile terminal. This characteristic has significant implications for Connected Vehicle applications and devices, since it requires that the mobile terminal actively request information as opposed to passively receiving it by virtue of being within radio range of a DSRC RSU or OBE. Since the mobile terminal doesn't know if any given section of road includes potential hazards, it must regularly request this data, and most requests will typically result in a "no data" response. While this approach is somewhat inefficient, the requests and "no-data" responses are expected to be small (in terms of message size), so the data load imposed by these null transactions is expected to be well within the capability of the system.

Last, latency requirements must be considered as part of the performance requirements of the communications technology. AASHTO describes three of the more important types of applications and their requirements:

- **Roadside Hazard Warnings/Alerts**: In general, roadside warnings and alerts persist on the roadway for some time, and it similarly takes some time simply to identify hazards and to generate alert information. As a result, the maximum latency requirement for this sort of application presumably ranges between minutes and tens of minutes.

- **Probe Data Collection**: Maximum latency requirements for probe data relate to how old the reported data can be. Since probe data may be collected in the vehicle and sent when the vehicle has travelled some distance, there will be some inherent latency in all probe data. As a lower bound, at an average speed of 30 mph, a vehicle reporting probe data every two miles will generate a data latency of four minutes from the time the first data element is stored to when the full set of data is transmitted. The subsequent communications latency in reporting the probe data to a center system could then reasonably be within 10-30 seconds.

- **Time Sensitive Applications**: The most obvious time critical V2I application is intersection safety. This is because the signals are changing state over time. The signal timing does not change arbitrarily. The key requirement for SPaT applications is that the SPaT data a vehicle uses at the time it reaches the dilemma zone must be valid until that vehicle passes through the intersection or stops. The tolerable latency for a BSM is actually the time interval in which the vehicle trajectory may change sufficiently that the projected position of the vehicle would be outside the tolerable position error for the application. For locally-transmitted (e.g., DSRC) systems the latency will be driven by the message repeat interval and the communications latency. For request-based (e.g., cellular) systems the latency will be the server response and communication medium propagation time. For either of these cases the latency is non-zero, and this latency will need to be added to the yellow-all-red cycle to assure safety. Obviously lower latency systems will have lower impact on the signal timing.

AASHTO's Alternative Communications Technologies Analysis

AASHTO performed analysis on Wi-Fi, a well-known communications system that is used for wireless Ethernet connections for PCs, phones, and other personal computing devices. Notably, Wi-Fi standards depend only on an IP protocol, and it is assumed that the communicating devices have an IP address. As part of forming an IP-based network, these standards include a network association process. This process essentially enables a terminal to "join" (or associate with) the network. In that process the terminal tunes to the channel that the network is using, learns the addresses of all other nodes on the network, and has its address distributed to all other nodes on the network. This process is relatively slow and consequently, the networks are configured to support fixed or slow-moving terminals. The association process is too slow to enable a moving vehicle to reliably attach to the network (associate) and communicate data. In addition, vehicles in a roadway environment will be entering and leaving the radio footprint (the coverage area) of the network at a rapid rate. This means that the network is constantly changing, and the network management function (typically performed by the base station) would need to constantly change the network definition (which terminals and their addresses are part of the network).

While DSRC uses the Wi-Fi protocol, it has specifically eliminated this network association aspect. Wi-Fi systems generally operate with a maximum outdoor range between 140 and 250 meters, which is acceptable for V2I and I2V communications, but, in general, the systems actually achieve much shorter ranges.

AASHTO also analyzed SDARS and noted that while the system offers nationwide coverage and thus could be useful for some I2V applications, it is not capable of V2I communications since it has no user terminal uplink (e.g. vehicle to satellite). It is also relatively low bandwidth. The entire channel capacity of a typical SDARS satellite is 4 Mbps, and, because it is a subscription service, most of this data capacity is allocated to subscriber-based audio channels. And, using it on a large scale could quickly become very expensive. (Subscriber revenue for the entire SDARS system is about $2.5 B per year; this equates to about $0.16 per KB. This is roughly 10 times that of cellular.) In addition there are uncertain delays between when data is submitted to the SDARS system and when that data is actually broadcast. For some applications this latency and the uncertainty in the latency would not be acceptable for a number of applications.

Summary

AASHTO determined that the cellular approach would be feasible for many non-safety applications, although latency remains a concern for the crash avoidance safety applications (both V2V and V2I). For most V2I applications that are not time-sensitive, the cellular approach is feasible, and it may represent a faster adoption, lower cost, and significantly lower risk option than DSRC.

- Faster adoption may result due to the existence of hundreds of millions of smartphones already in the field, many of which could access Connected Vehicle services today with the simple installation of a (presumably free) application. This means Connected Vehicle services could be rolled out immediately rather than waiting decades for the equipped vehicle population to grow. In addition, the smart phone turnover rate is about 200 million units every 18 months, or 133 million per year—about 10 times the turnover of vehicles, so even if new hardware features were required the rollout speed would be substantially faster.

- Lower cost is partly a result of lower user terminal cost—most people use phones for other applications, so the incremental cost of Connected Vehicle applications is nearly zero—and substantially reduced infrastructure costs. The primary communications infrastructure is already in place and is funded by the user's subscriptions. To support a cellular-based system the jurisdictional stakeholders (the presumed infrastructure deployers) would need only to implement appropriate back office systems.

- Lower risk arises from the fact that the cellular user base already exists. This may be true if the infrastructure rollout lags, and/or if the limited number of equipped vehicles in the early years causes consumers to feel the system provides little real value. Lower cost solutions that provide benefits more quickly are also less likely to be subject to the uncertainty of political sensibilities.

Cellular Implementation Examples

As part of their analysis, AASHTO provides examples of typical, existing cellular services implementations. Some of these are embedded connected vehicle systems, and others are currently based on portable/handheld terminals, but could just as easily be embedded in the vehicle.

Probe Data Collection

Probe data typically consists of operational data collected from moving vehicles. This data may be very simple, such as position and time, or it may include a variety of vehicle operational parameters such as ABS events, speeds, accelerations, etc. Several service providers collect time and position data from location aware cellular phones. These data are typically provided under a privacy agreement that essentially allows the consumer to receive location oriented services in exchange for providing their position at regular intervals. The collected data is generally aggregated and anonymized, and then reused for other purposes, for example as traffic congestion information described next.

Traffic Information Distribution

A variety of other cellular-based "connected traveler" systems are available. For instance, a number of private sector firms provide traffic information services based on the collection of cell phone location data. These traffic services highlight congestion by color coding speed levels relative to free flow speed on any streets where cellular data from phones has been collected—surface streets as well as major arterials and freeways.

Speed Harmonization and Green Wave

Several companies have leveraged smartphones to provide information to drivers relating to traffic signals. One example is imagreendriver.com, headquartered in Portland, OR. This system uses real-time traffic signal data provided by the city to support routing and "green wave" applications on smartphones. Using the traffic signal data, one application will determine the optimal route that assures the shortest wait at traffic signals. Another application shows the driver the current state of an approaching traffic signal, and suggests an optimal speed to drive to assure that the light will be green when the vehicle arrives at the intersection. As such, this serves as an example of a time-sensitive V2I application based on cellular communications.

B.II.C. 2013-2014: For Backhaul Services and Applications by AASHTO[110]

Conventional connected vehicle architectures assume that field equipment and center facilities are connected by a communications link. This is typically called a "Backhaul Network". In these systems the Center can send information to field terminals (e.g. messages to be transmitted by the field terminal) and the field equipment can send information back to the center facility. The information sent back to the center facility may be status information about the field terminal, or it may be local information relating to other field equipment such as signal controllers that are attached to the field terminal. It may also be information received from nearby mobile terminals and forwarded to the center by the field terminal. Some connected vehicle architectures may not use field equipment. In this case communications between the mobile terminals and the center facilities would be over a wide area network.

Numerous technologies can be used to provide backhaul communications. AASHTO performed an analysis of the typical applications that a DSRC RSU might support, and the expected data loads that each application would impose on the backhaul. For most applications, the overall load is not particularly high. Key exceptions are the distribution of CRLs, the capture of Basic Safety Messages (BSMs) at DSRC RSUs, and private services.

The highest data load imposed by a broadcast application (e.g., safety messages being broadcast from the DSRC RSU) is posed by roadway alert messages (72Kbps). Signal Phase and Timing (SPaT) also imposes a modestly high data rate requirement (42.8 Kbps), with the assumption that the SPaT is generated centrally and then sent to a DSRC RSU over the backhaul. When generated locally, the data load is not imposed on the backhaul.

AASHTO analyzed these needs over 14 different types of communications technologies:

- Wide Area (WAN)
  - Cellular (LTE)
  - Cellular (General Packet Radio Service; GPRS)
  - WiMAX
  - FSS
  - SDARS
- Local Area Network (LAN)
  - Ultra-Wideband (UWB)
  - Wi-Fi
  - DSRC

---

[110] The following section is summarized from the draft *National Connected Vehicle Field Infrastructure Footprint Analysis: Deployment Concepts*, September 2013, pages 63-70, at http://stsmo.transportation.org/Documents/Task%206a%20AASHTO_CV_Footprint_Deployment_Scenarios_v2.pdf. (Last accessed June 2015)

- o   ZigBee
- Point to Point
  - o   Fiber
  - o   Digital Subscriber Line (DSL)
  - o   Cable TV
  - o   Microwave
  - o   Plastic Leaded Chip Carrier (PLCC)

In summary, AASHTO concluded that the non-satellite wide area network technologies (WAN) are able to support all applications and data loads.

Appendix C   DSRC in Use Today

Appendix C describes today's DSRC test beds for research and real-world operations. Furthermore, this appendix also provides the planned DSRC uses. The appendix is structured as follows: Research Test Beds Using DSRC, Operational Uses of DSRC and Planned Uses of DSRC.

C.I.   Research Test Beds using DSRC

C.I.A.   California

In 2005, the Palo Alto, CA test bed was established by the California Department of Transportation (Caltrans) and the California Partners of Advanced Transit and Highways (PATH), part of the University of California – Berkeley's Institute of Transportation Studies. To assess real-world implementations of Connected Vehicle infrastructure, architecture and operations[111], the DSRC test location was designed on a 6-mile segment of the El Camino Real. However, before advanced testing could occur, funding fell through to develop more advanced equipment. In 2007, the Metropolitan Transportation Commission partnered with Caltrans and PATH to expand the original programs of the test bed oriented to relieve congestion and to demonstrate the capabilities and feasibility of Connected Vehicle technologies.[112] In late 2008, the test bed was one of two participating in the SafeTrip-21 (Safe and Efficient Travel Through Innovation and Partnerships in the 21st Century) program. The program utilized the 27 signaled intersections, and established 5.9 GHz DSRC roadside infrastructure to improve safety, mobility, energy independence and environmental stewardship.[113]

Since development, the test bed has housed many other DSRC research projects including: PATH test intersection, at-grade light rail crossings, station traveler information, cooperative intersection collision avoidance systems, signalized left turn assistance and lane-level positioning capability.[114] To review and enhance these applications Caltrans has developed regular partnerships with local automotive OEMs and established integration with the national data backhaul system. In 2012, to maintain the status of a leading Connected Vehicle test bed, the site updated its technology: replacing the original 5.9 GHz DSRC radios with updated

---

[111] http://www.ite.org/meetings/2012TC/Session%2016_Greg%20Larson.pdf slide 3.
[112] http://www.michigan.gov/documents/mdot/09-12-2013_International_Survey_of_Best_Practices_in_ITS_434162_7.pdf page 15.
[113] ITS Research Results: ITS Program Plan 2008, page 51 (located at: http://ntl.bts.gov/lib/30000/30800/30867/ITS_Research_Results_ITS_Program_Plan_2008_-_ITS_Report.pdf).
[114] http://www.ite.org/meetings/2012TC/Session%2016_Greg%20Larson.pdf slide 2.

---

equipment consistent with the infrastructure found in Ann Arbor, Michigan for the Safety Pilot Model Deployment.[115] During the upgrade, Palo Alto demonstrated that, if roadside equipment exists, most connected software and infrastructure can be re-purposed and re-used to comply with the needs of updated DSRC equipment.

The Palo Alto, California test bed was developed by Caltrans with the cooperation of several vehicle manufacturers and USDOT to assess real-world prototypes of vehicle-infrastructure integration. The experimental test bed itself consists of several 5.9 GHz DSRC RSUs deployed along a two-mile stretch of the El Camino Real (CA-82) at 11 consecutive signalized intersections in Palo Alto, integrated with signal controllers and the *511* traveler information center in Oakland. The site has been and will be used for multiple application tests including: traveler information, electronic payments, Cooperative Intersection Collision Avoidance Systems (CICAS), transit information, eco-driving, and the Multi-modal Intelligent Traffic Signal System (MMITSS).

C.I.B. Michigan

Since 2008, the Department has upgraded the Michigan test bed as its major investment and made it widely available for use in testing applications, services, and components. The test bed is located in Oakland County, Michigan, centered in the cities of Novi, Farmington, and Farmington Hills with recent expansion into Southfield. At this location, the test bed offers 5.9 GHz DSRC RSU installations, back office servers and support, on-board equipment, as well as the SPaT and Geometric Intersection Design (GID) applications. It also uses the latest technology standards and architecture and is based on the systems engineering principles in the Core Systems project.[116] Older California and New York sites that were developed for Vehicle- Infrastructure Integration (VII) tests are now upgraded and connected so that multiple sites function as one system. When operational, the Safety Pilot Ann Arbor, Michigan site will be connected as part of this system as well.

C.I.C. New York State

The New York Test Bed was created in 2008 to debut at the 2008 ITS World Congress.[117] During the World Congress, the test bed was most known for its extensive connected infrastructure, testing area, and simultaneous application testing. With 42 miles of testing corridor housing more than 40-5.9 GHz DSRC RSUs and the ability to demonstrate more than 20 applications (e.g. travel time information, dynamic message sign (DMS) messages,

---

[115] http://ntl.bts.gov/lib/51000/51000/51002/DF3F1F7.pdf page 25.
[116] http://www.its.dot.gov/research/systems_engineering.htm
[117] http://i95coalition.org/i95/Portals/0/Public_Files/CVII/Presentations/NYS%20Existing%20VII%20-%20May%202009.pdf slide 5

emissions calculations, intersection safety, transit priority, multimodal information, Connected Vehicle probe data, work zone safety warning, warning sign enhancement, curve warning, commercial vehicle routing information, and vehicle restrictions),[118] the test bed was the second largest Connected Vehicle test bed in the country.

Although no longer one of the largest in the country, the test bed is still active. Most recently, the New York test bed is most known for the deployment of the Commercial Vehicle Infrastructure Integration (CVII) program. This three-year program utilized 31-5.9 GHz DSRC RSUs over a total of three testing corridors: 13 miles of I-87, 42 miles of I-495 and an unspecified segment of I-90. In 2011, phase 1 of a potential three phased commercial vehicle infrastructure program was completed. This first phase demonstrated 5.9 GHz DSRC roadside and on-board equipment communications in commercial vehicles including: integrating commercial vehicle probe data with traveler information, driver credential verification, wireless roadside inspection, and V2V communications. Phase 2 was completed in 2012 focusing on developing Connected Vehicle applications: blind spot warnings, hard braking events, tailgate warnings, unsafe-to-merge/pass scenarios and a railroad crossing grade warning system.[119]

C.I.D.   Orlando, Florida[120]

Deployed in 2011, the Orlando Test Bed is the only transportation management center-based test bed operational in the country. The test bed is controlled by a transportation data center that aggregates and disseminates information. The 25 mile testing site hosts 26-5.9 GHz DSRC RSUs to enhance their already largely successful SunGuide program, an advanced transportation management software. The addition of the RSUs expands the software to provide basic safety messages and receive traveler information. Although created for the World Congress, the Florida DOT has maintained the software and the RSUs and continues using the applications. After the success of the testing, the lessons learned created by the developers of the program became a national standard in developing new test beds.

The Orlando test bed was developed by Florida DOT as a test site for the 18th ITS World Congress in 2011.[121] The test bed infrastructure consists of RSUs along freeway and arterial roadways connected to Florida DOT SunGuide servers and data management systems over their fiber network. The system was used to test capturing Basic Safety Messages from vehicles operating on the test bed to calculate travel times.

---

[118] http://www.michigan.gov/documents/mdot/09-12-2013_International_Survey_of_Best_Practices_in_ITS_434162_7.pdf page 23

[119] http://www.swri.org/4org/d10/isd/ivs/CVII.htm

[120] http://www.dot.state.fl.us/trafficoperations/its/projects_deploy/cv/connected_vehicles-wc.shtm

[121] http://www.dot.state.fl.us/trafficoperations/its/projects_deploy/cv/connected_vehicles-wc.shtm, accessed February 2014.

With recent discussions about connected automated vehicle research, the Florida DOT has shifted gears slightly; however, it was stated that the test bed may be used as an outlet to combine DSRC RSUs and Connected Vehicle technologies with automated vehicles to complement automated vehicle technologies.[122]

In 2015, the Integrated V2I Prototype was installed as part of the Orlando test bed. Testing will continue in 2016, leading to development of the V2I Reference Implementation in the 2016-2017 timeframe.

C.I.E. Arizona

The Maricopa County Department of Transportation (MCDOT) SMARTDrive Program test bed in Anthem, Arizona includes six pole-mounted 5.9 GHz DSRC roadside units integrated with traffic signal controllers at signalized intersections. The test bed was initially developed in 2011 to advance multiple vehicle signal priority technologies in a real-world traffic environment.[123] The units are installed along a 2.3-mile segment of Daisy Mountain Drive in Anthem, Arizona, just north of Phoenix.

DSRC was selected as the preferred communications medium largely because it provides the low latency performance needed for a number of the applications investigated at the test bed. Other considerations that led to the selection of DSRC include that it is licensed and therefore less likely to suffer from interference; it is secure; and it can be coordinated with USDOT and other DSRC testing.[124]

The MCDOT SMARTDrive Program test bed expanded on previous Arizona-based traffic signal research (Next Generation of Smart Traffic Signals in 2007 and Arizona E-VII program in 2008[125]). The MCDOT SMARTDrive test bed is currently being used to field test Multimodal Intelligent Traffic Signal Systems (MMITSS) capabilities. The testing is sponsored by the Cooperative Transportation Systems Pooled Fund Study in partnership with USDOT. The test bed is located on public streets but is utilized for controlled testing and is not operational day-to-day.

MMITSS is one of the Connected Vehicle Dynamic Mobility Application high-priority bundles identified through stakeholder engagement. The purpose of the MMITSS applications is to integrate new information from connected travelers and existing information from infrastructure

---

[122] http://floridaits.com/ConnVeh.html (last accessed July 2015)

[123] http://www.sonorannews.com/archives/2011/111116/frontpage-smartdrive.html (last accessed July 2015)

[124] Faisal Saleem, ITS Branch Manager and SMARTDrive™ Manager, Maricopa County Department of Transportation; interviewed by Matt Burt (USDOT Volpe Center); June 27, 2014.

[125] http://www.michigan.gov/documents/mdot/09-12-2013_International_Survey_of_Best_Practices_in_ITS_434162_7.pdf. (last accessed July 2015)

based detection systems into a safer and more effective traffic signal control system. This integrated information can be used to make improvements in traffic control algorithms and logic resulting in better performing and safer operating systems.[126] The MMITSS application bundle comprises applications pertaining to overall traffic flow; freight, emergency and transit vehicle priority; and pedestrian mobility.

MMITSS testing comparable to that being done in the MCDOT SMARTDrive Program test bed in Arizona is being done concurrently in Palo Alto, California. The California testing is exploring interoperability issues by implementing similar MMITSS capabilities in conjunction with different devices and standards as those used in the MCDOT SMARTDrive test bed.

The MCDOT is planning to expand the test bed through two related projects that include additional 5.9 GHz DSRC RSUs at seven or eight signalized intersections contiguous to the existing test bed. One of the projects will be implemented in conjunction with the Arizona Department of Transportation (ADOT) and includes adding 5.9 GHz DSRC units at the signalized intersections of an arterial street with ADOT-operated freeway ramps.

MCDOT also plans to apply the findings of their testing to implement MMITSS capabilities in other locations in a fully-operational mode. These plans include pedestrian mobility capabilities at a signalized crosswalk in Sun City, Arizona and freight traffic signal priority capabilities on County Road MC-85, a freight-intensive corridor.[127]

MCDOT concludes that testing has demonstrated a number of important DSRC capabilities. These capabilities include reliability (including successful operation through two Arizona summers), low latency V2I communication effectiveness, and successful interface with traffic signal controllers using NTCIP and the SAE J2735 standard.[128]

C.I.F. Ann Arbor, Michigan

Ann Arbor houses the largest test bed, to date, for DSRC Connected Vehicle application programs.[129] The site is a combined effort from UMTRI and USDOT. It contains 73 miles of roadway, 29-5.9 GHz DSRC RSU installations, and over 2,800 designated testing vehicles.[130]

---

[126] University of Arizona (lead); *MMITSS Final Concept of Operations*; Version 3.1; December 4, 2012. http://www.cts.virginia.edu/wp-content/uploads/2014/05/Task2.3._CONOPS_6_Final_Revised.pdf (last accessed July 2015)

[127] Saleem interview; June 27, 2014.

[128] Saleem interview; June 27, 2014.

[129] http://www.its.dot.gov/newsletter/august2012.htm (last accessed July 2015)

[130] Download at: http://www.google.com/url?sa=t&rct=j&q=&esrc=s&source=web&cd=8&ved=0CFYQFjAH&url=http%3A%2F%2Fwww.transportation.org%2FDocuments%2F102013%2520Steudle_Innovation_CV%2520Update%2520AASHTO_nn.pptx

---

The test bed is most known for its year-long Safety Pilot Model Deployment Program. The purpose of the program is to test system performance, evaluate human factors and usability, and observe policies and processes and collect empirical data to present a more accurate, detailed understanding of the potential safety benefits of Connected Vehicle technologies.[131]

Although the program results are still being reviewed, initial statements indicate that 90 percent of tested respondents felt positively about the safety applications they experienced and feel the overall benefits of the technology far outweigh the potential drawbacks. Despite critiques of distraction, the study also found that 74.5 percent of respondents indicated that the applications were no more distracting than listening to the radio.[132]

USDOT will continue to maintain, operate and upgrade the Safety Pilot test environment in Ann Arbor, Michigan through a cooperative agreement with UMTRI. The test site will continue to be used by USDOT and its external partners for research and deployment of Connected Vehicle technology.

C.I.G. Northern Virginia

The state of Virginia has two separate Connected Vehicle test beds. The Virginia Smart Road facility opened in 2012[133] and consists of a 2.2-mile controlled-access roadway[134] (not open to general traffic) built to Federal Highway Administration standards. The Smart Road facility is located in the southern portion of the state, in Blacksburg, Virginia. The facility is owned and maintained by the Virginia Department of Transportation (VDOT) and managed by the Virginia Tech Transportation Institute (VTTI). The Smart Road facility includes approximately 10-5.9 GHz DSRC RSUs[135] and a fleet of approximately 12 vehicles of various types, including sedans, sport utility vehicles, a motor coach, motorcycles, and a semi-tractor trailer.[136]

The other Connected Vehicle test bed is located in the northern part of the state in Fairfax County, Virginia, and opened in 2013.[137] The test bed is a VDOT facility developed in

---

&ei=_zqQU9jKI6HgsATv0YJI&usg=AFQjCNHTTNQLC69MiT3H0zYYnOMmNM-LzA&bvm=bv.68235269,d.cWc slide 5. (Last accessed July 2015)

[131] http://www.its.dot.gov/safety_pilot/index.htm (last accessed July 2015)

[132] http://www.its.dot.gov/newsletter/august2012.htm (last accessed July 2015)

[133] http://www.cvi-utc.org/?q=node/26 (last accessed July 2015)

[134] Virginia Tech Transportation Institute; *The Future is Now: Active. Connected. Automated.; January 2014*; http://issuu.com/vtti/docs/flipbook_web_final_jan_2014/37?e=0/6276481 (last accessed July 2015)

[135] Cathy McGhee, Associate Director for Safety, Operations and Traffic Engineering, Virginia Department of Transportation; interviewed on July 2, 2014 by Matt Burt, USDOT Volpe Center.

[136] Virginia Tech Transportation Institute; *The Future is Now: Active. Connected. Automated.; January 2014*; http://issuu.com/vtti/docs/flipbook_web_final_jan_2014/37?e=0/6276481 (last accessed July 2015)

[137] Virginia Tech Transportation Institute; *The Future is Now: Active. Connected. Automated.; January 2014*; http://issuu.com/vtti/docs/flipbook_web_final_jan_2014/37?e=0/6276481 (last accessed July 2015)

---

partnership with VTTI, the University of Virginia, and Morgan State University as part of the Connected Vehicle/Infrastructure University Transportation Center funded by USDOT. The test bed consists of 455.9 GHz DSRC RSUs deployed along I-66 and the parallel Routes 29 and 50.

Often, the controlled-access Smart Road facility is used for initial, smaller-scale testing; followed by larger-scale, live traffic operations in the Fairfax Test Bed. The two test beds have been used for a wide variety of safety, mobility, and environmental applications on both freeways and arterial roadways. VDOT research is particularly focused on V2I and I2V applications.[138]

VDOT and its partners are planning a major expansion of the Fairfax Test Bed in 2015 to support further investigation of communications system loading and latency issues under heavier traffic conditions.[139] The expansion includes recruitment of general public drivers and equipping 3,000 to 4,000 vehicles with DSRC and other technologies. The Fairfax Test Bed expansion will include an additional 25-5.9 GHz RSUs and will expand the test bed to a total of 25 miles.[140] Objectives of the expanded testing include investigation of: 1) System performance, including latency, under heavier vehicular traffic and communications loads; 2) Benefits associated with a number of VDOT high-priority applications, including traffic signals, pedestrian safety, work zone safety and mobility, and weather-related traffic advisories; and 3) Public acceptance and driver response.[141]

VDOT considers DSRC critical for traffic signal and other applications where low latency is necessary, with cellular a potential candidate for other applications and as a means to fill in coverage in areas that are not covered by DSRC. VDOT believes the unavailability of cellular service during times of extremely high usage is another reason to avoid relying entirely on cellular communications, citing their experience with a recent earthquake when the cellular system overloaded.[142]

C.I.H.   Canada

In its 2003 Report and Order, the FCC discussed the requirements for terrestrial licenses along the borders with Canada and Mexico, but noted that no international agreements exist concerning the current ITS allocation of the 5.9 GHz DSRC band spectrum. As a result, the FCC

---

[138] Cathy McGhee, Associate Director for Safety, Operations and Traffic Engineering, Virginia Department of Transportation; interviewed on July 2, 2014 by Matt Burt, USDOT Volpe Center.

[139] Cathy McGhee, Associate Director for Safety, Operations and Traffic Engineering, Virginia Department of Transportation; interviewed on July 2, 2014 by Matt Burt, USDOT Volpe Center.

[140] Virginia Department of Transportation, presentation, *Virginia Connected Test Bed*; presentation, May 28, 2014

[141] Cathy McGhee, Associate Director for Safety, Operations and Traffic Engineering, Virginia Department of Transportation; interviewed on July 2, 2014 by Matt Burt, USDOT Volpe Center.

[142] Cathy McGhee, Associate Director for Safety, Operations and Traffic Engineering, Virginia Department of Transportation; interviewed on July 2, 2014 by Matt Burt, USDOT Volpe Center.

"...proposed to apply the same technical restrictions at the border that we adopt for operation between service areas, i.e., operations must not cause harmful interference across the borders."

USDOT and Transport Canada, acting under the Regulatory Cooperation Council (RCC)[143], agreed in May of 2014 to perform a technical and policy comparison of the uses of the 5.9 GHz DSRC band to understand whether differences exist. This comparison will be available in 2015. In Canada there are two major projects underway to build and operate a network of five Connected Vehicle systems test beds that will provide input to the DSRC comparison analysis. The projects are:[144]

University of Alberta ACTIVE Test Bed. The Alberta Cooperative Transportation Infrastructure & Vehicular Environment (ACTIVE) project consists of two on-road test beds (arterial and freeway) and one laboratory test bed located in Edmonton, Alberta that will be owned and operated by the University of Alberta. The ACTIVE test beds will provide stakeholders with real-world environments for testing with an emphasis on data for active traffic and demand management, as well as providing real-time traffic density, flow and congestion data to support provincial/municipal traffic management.

University of British Columbia AURORA Test Bed. The University of British Columbia (UBC) Automotive Test Bed for Reconfigurable and Optimized Radio Access (AURORA) project consists of one on-road test bed (around the UBC campus) and one laboratory test bed in Vancouver, British Columbia that will be owned and operated by the UBC. AURORA will provide stakeholder access to the next-generation of Connected Vehicle systems, and wired and wireless communications technologies and resources to test and evaluate new and innovative products and services with an emphasis on commercialization and evaluation of technologies for wireless freight security and efficiency.

The key infrastructure for the on-road test beds will consist of equipment installed along the roadways including sensors, wired and wireless communications technology, and cameras that can monitor and talk with the various probe vehicles that will travel along the test beds' roadways. The key infrastructure for the on-campus labs will consist of facilities, information and communication technologies (ICT), and software to control and manage the on-road test beds.

---

[143] A government-to-government initiative started by the President (US) and the Prime Minister (Canada).
[144] University of Alberta, The Centre for Smart Transportation; *The ACTIVE-AURORA Test Bed Network*; described in: http://ntl.bts.gov/lib/52000/52600/52602/FHWA-JPO-14-125_v2.pdf p.95. (Last accessed June 2015)

C.I.I. Affiliated Connected Vehicle Test Beds

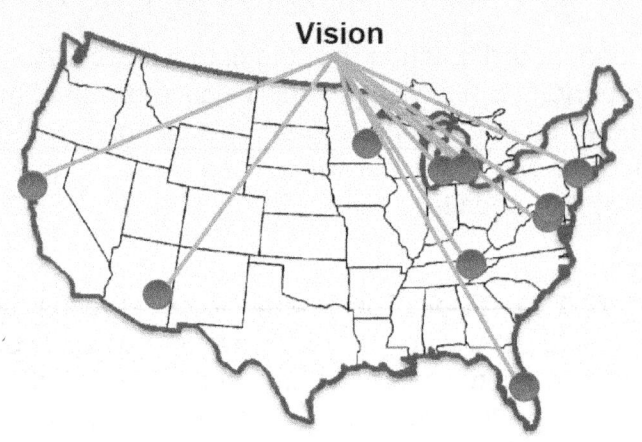

Figure C-1: Test Bed Sites

Image: USDOT

Connected Vehicle Test Beds are real-world, operational test beds that offer the supporting vehicles, infrastructure, and equipment to serve the needs of public- and private-sector testing and certification activities. Today's current test environments (described in C.I.A-C.I.H) support continued research and testing of connected transportation system concepts, standards, applications, and innovative products. Test environments also serve as a precursor or foundation for state and local deployments of wireless communication between vehicles. They are expected to generate sustainable markets for the private sector, as the test environment will enable products and applications that will deliver benefits to state and local consumers who purchase them.

In supporting these test beds, USDOT has enabled multiple, interoperable locations as part of a connected system moving toward nationwide deployment. The sites, along with over 70 partners (many of whom are from the private sector), share information and lessons learned and participate using:

- Common architecture
- Common standards
- Independent operations
- Shared resources (for instance, security)

C.II. Operational Uses of DSRC

C.II.A. Southeastern Michigan 2014 Project

In 2013, the Southeast Michigan test bed was upgraded with 50-5.9 GHz DSRC RSUs to allow testing by industry; and for USDOT to pursue new concepts for deployment (e.g., new data exchange or warehouse strategies). This test bed was originally created in 2007. Current operations use all seven channels to test V2I and I2V data flows and a mix of safety and non-safety data flows. Most notably, USDOT plans to use this project site to pioneer many essential Connected Vehicle environment capabilities, such as Security Credential Management systems (SCMS) and interoperability testing, during the estimated two year timeline. This real-world, operational test bed will further support development of certification processes and policies.

C.II.B. San Francisco Municipal Transit Authority[145]

Developed in 2012, the San Francisco test bed has been working on plans to integrate Connected Vehicle technologies over the 5.9 GHz DSRC wireless broadband system, adding to their other systems, such as their 4.9GHz safety system. The test bed is currently investigating the use of 5.9 GHz DSRC equipment as a replacement for the current Land Mobile Radio system, creating an open infrastructure system to support a high occupancy toll (HOT) lane tolling application, data transfers, and transit signal priority.

C.II.C. Seattle/King County

King County Metro is in the project-development stage (budget and scope) of a major revision to their ITS Architecture. DSRC will be part of that architecture, and its role will be determined during the development of the Concept of Operations. At a minimum, King County Metro is planning to use DSRC for Connected Vehicle safety applications; in particular, the agency has developed a concept for pedestrian-bus interactions to give the bus a protected right turn at heavy pedestrian intersections.

The agency also believes that Transit Signal Priority will be performed using DSRC where it's available.[146] Currently, the agency's existing communications network is deployed on six Bus Rapid Transit corridors covering over 100 miles, with over 200 intersections. The network supports several systems including transit signal priority (TSP), AVL bus location data, passenger information signs, and electronic fare payments. King County DOT uses the wired portion of the network for signal controller interconnect.

The network uses the 4.9 GHz public safety band. However, King County Metro expects the devices and technologies associated with this band to be phased out over the next years and will move to the 5.9 GHz DSRC band. In anticipating changes to the market, the agency developed a transition path from 4.9 GHz to 5.9 GHz in 2011. Their analysis noted the consistencies associated with the standards and further noted that at the time of implementation

---

[145] http://www.apta.com/mc/its/program/Documents/San-Franciscos-Implementation-of-DSRC-5.9-GHz-Daniel-Buckley.pdf (last accessed July 2015)

[146] The agency's current transit signal priority architecture supports concurrent use of any of the five TCIP/NTCIP 1211 standard scenarios.

technologies using DSRC were not available. Working with the architecture designers, the agency incorporated the capability for migration to 5.9 GHz DSRC with market and equipment maturity. Because their network is based on IP and Transit ITS standards, it will also support a wide range of interoperable unified communications extensions such as voice, video, and multi-media apps.[147]

C.III. Planned Uses for DSRC

C.III.A. Connected Vehicle Pilot Deployments

Based on successful results of the Connected Vehicle research program, and the recent decision by NHTSA to pursue vehicle-to-vehicle communications safety technology for light vehicles, a robust Connected Vehicle pilots program is envisioned as a mechanism to spur the implementation of Connected Vehicle technology. These pilots will serve as initial implementations of V2V and V2I technology deployed in real world settings with the aim of delivering near-term safety, mobility, and environmental benefits to the public.

The program seeks to spur innovation among early adopters of connected vehicle application concepts, using best available and emerging technologies. Pilot deployments are expected to integrate connected vehicle research concepts into practical and effective elements, enhancing existing operational capabilities. The intent of these pilot deployments is to encourage partnerships of multiple stakeholders (e.g., private companies, States, transit agencies, commercial vehicle operators, and freight shippers) to deploy applications utilizing data captured from multiple sources (e.g., vehicles, mobile devices, and infrastructure) across all elements of the surface transportation system (i.e., transit, freeway, arterial, parking facilities, and toll ways) to support improved system performance and enhanced performance-based management. The pilot deployments are also expected to support an impact assessment and evaluation effort that will inform a broader cost-benefit assessment of Connected Vehicle concepts and technologies.

Program Applications

Pilot deployments are expected to build upon the USDOT-sponsored application and technology research. Prototypes of selected Connected Vehicle applications are currently under development and testing, with pilot deployment planned to follow. Some concepts of operation, system requirements, and design documents will be made available, as well as algorithms and source code associated with these prototypes. A pilot deployment concept need not include all

---

[147] *Connected Vehicle at 4.9GHz for Transit ITS: Unified Communication Architecture for an ITS Enterprise and Pathway to DSRC*, presentation and paper by Bryan Nace, CCNP–DKS Associates and John Toone, MPA–King County Metro, October 19, 2011. Located at: http://itswc.confex.com/itswc/WC2011/webprogram/Paper1868.html. (last accessed July 2015)

of the specific technologies identified in the Connected Vehicle research effort. However, each pilot deployment should combine concepts from multiple USDOT application development efforts.

These pilot deployments will also serve as a precursor or foundation for state and local deployments of wireless communication between vehicles. They are expected to generate sustainable markets for the private sector, as the test environment will enable products and applications that will deliver benefits to state and local consumers who purchase them.[148]

C.III.B. European Union: Rotterdam to Vienna Corridor

One of the front runner ITS activities in Europe is the trilateral Cooperative ITS corridor between Vienna, Austria, Munich and Frankfurt in Germany, and Rotterdam in the Netherlands. Deployment of the corridor is organized and managed by the transport ministries and automotive industry of the three countries. The roadside Cooperative ITS infrastructure for the initial services along the corridor are planned to be installed during 2015 and will become operational as installations occur. The result is that vehicle manufacturers can expect day-one applications and services to be available to the market. Similar to the US, European OEMs will need to plan for changing their manufacturing process to integrate vehicle-based components.

The deployment is governed by a European Union Member States Memorandum of Understanding with industry groups under similar agreements. It will serve as a V2I-based roadwork warning information system for the corridor connecting the cities in the three countries. Communication from vehicles to infrastructure will be established via short range communication (Wi-Fi, DSRC) or cellular (3G, 4G). Initial applications are expected to increase road safety and provide the basis for improved traffic flow. It is anticipated that the success of this project will lead to future initiatives that will connect to the Austrian-German-Dutch Cooperative ITS Corridor, expanding the overall system. It is also expected that connections to the urban network will evolve as current cooperative ITS sites also cover parts of the urban network (e.g. Helmond, Frankfurt, Vienna).[149,150]

---

[148]http://www.its.dot.gov/pilots/index.htm (Last accessed June 2015)

[149] Cooperative ITS Corridor Joint Deployment. July 16, 2014 (accessed). *Cooperative ITS Corridor Joint Deployment Fact Sheet.* Ministry of Infrastructure and the Environment of the Netherlands; Ministry of Transport and Digital Infrastructure of Germany; and Federal Ministry for Transport, Innovation and Technology of Austria.

[150] Amsterdam Group. June 7, 2013. *Draft Roadmap between automotive industry and infrastructure organizations on initial deployment of Cooperative ITS in Europe Version 1.0.*

Appendix D History of DSRC: Policy and Technical

Appendix D describes the background and context for assessing the status of the DSRC technology and applications developed through research and development. The chapter is divided into two sections:

I. Section D.I provides an historical overview of the policy milestones from the inception of the Federal Communications Commission's (FCC) rule to the present; and identifies future, anticipated milestones.

II. Section D.II provides a similar, historical overview of the technical research and development of the communications requirements, applications, and standards that support successful DSRC operations.

D.I. DSRC: Policy History

This section offers a summary of the key policy milestones in developing DSRC for operational use. The major DSRC policy milestones are summarized, as follows:

DSRC Policy Milestones
FCC Rule:
- 1998: FCC initiates rulemaking for allocating 75 MHz at 5.850-5.925
- 1999: FCC issues a Federal Register announcement regarding the allocation of spectrum for ITS

Spectrum Sharing Protocol:
- 2006: Spectrum Sharing Agreement with the Fixed Satellite Services (FSS) is filed with the FCC
- Service Rules:
    - 2003: FCC adopts a Report and Order (R&O) that establishes licensing and service rules
    - 2006: FCC issues a Memorandum, Opinion, and Order amending service rules
    - 2011: USDOT provides greater definition regarding use of Channel 172

DSRC Operations:
- 2012: USDOT provides FCC with information related to Safety Pilot Tests
- 2013: FCC issues an NPRM on sharing with unlicensed devices
- 2014: NHTSA issues a Decision to Move Forward with Vehicle-to-Vehicle Communication Technology for Light Vehicles

Upcoming Policy Milestones:
- Future: NHTSA to issue a decision on Heavy Vehicles
- 2015: FHWA to issue Guidance to State and Local Agencies on:
    - Licensing and Siting for DSRC Infrastructure
    - A Systems Engineering Approach to Connected Vehicle Communications

D.I.A.   1998-1999: History of the FCC Rule

In October 1999, the FCC announced the allocation of 75 MHz of spectrum for intelligent transportation services with the following statement:

> "The FCC decided to use the 5.850-5.925 GHz band for a variety of Dedicated Short Range Communications (DSRC) uses, such as traffic light control, traffic monitoring, travelers' alerts, automatic toll collection, traffic congestion detection, emergency vehicle signal preemption of traffic lights, and electronic inspection of moving trucks through data transmissions with roadside inspection facilities. ….
> The FCC said providing additional spectrum for ITS services would further the goals of Congress, the Department of Transportation and the ITS industry to improve the efficiency of the US transportation infrastructure and to facilitate the growth of the ITS industry."

The FCC initiated the rulemaking proceeding in 1998 and adopted the first R&O[151] on October 21, 1999 (which it released on October 22, 1999) allocating 75 MHz of spectrum at 5.850-5.925 GHz to the mobile service for use by DSRC systems operating in the ITS radio service. This R&O did not adopt licensing and service rules or spectrum channelization plans. The FCC announced another NPRM on November 7, 2002 that sought comment on licensing and service rules.[152] The FCC also requested comments on whether prior coordination between DSRC and Fixed Satellite Services would be needed and under what conditions.

The current R&O[153] was adopted on December 17, 2003 and released on February 10, 2004. This R&O established licensing and service rules for DSRC operating in the ITS service. It also adopted the interoperability standard know as ASTM E2213-02. It made possible licensing in the 5.9 GHz DSRC band for both public safety and nonpublic safety uses. In this current R&O, the FCC notes "...that DSRC system design is in its infancy and we expect further development and refinement."[154]

The FCC allocation of 5.9 GHz is not exclusive to ITS. There is a co-primary Federal Government radiolocation allocation (for use by high-powered military radiolocation) in the 5.850-5.925 GHz band and a co-primary FSS allocation. Amateur service has a secondary allocation in the band also. Industrial, Scientific and Medical (ISM) equipment may also operate in the 5.850-5.875 MHz portion of the band.

D.I.B. 2003-2006: History of Spectrum Sharing with the Satellite Industry

In support of the FCC requests regarding coordination between DSRC and FSS, a joint committee was convened in 2003 to determine the potential for interference between the two services and to develop guidelines and recommendations for siting of stations to reduce the potential for inter-service interference. Three issues have been extensively analyzed:

- In-band sharing;
- Adjacent band interference to DSRC from FSS operations; and
- The potential for aggregate interference from DSRC into FSS uplinks (i.e., increasing the "noise floor").

---

[151] This R&O is FCC 99-305 in ET Docket 98-95 and RM-9096 at: http://www.fcc.gov/oet/dockets/et98-95. (Last accessed June 2015)
[152] See, story titled "FCC Adopts Intelligent Transportation Systems NPRM" in TLJ Daily E-Mail Alert No. 546, November 11, 2002 at: http://www.techlawjournal.com/alert/2002/11/11.asp. (Last accessed June 2015)
[153] http://hraunfoss.fcc.gov/edocs_public/attachmatch/FCC-03-324A1.pdf. This R&O is FCC 03-324 in WT Docket No. 01-90, ET Docket 98-95 and RM-9096. (Last accessed June 2015)
[154] p. 18 http://hraunfoss.fcc.gov/edocs_public/attachmatch/FCC-03-324A1.pdf. (Last accessed June 2015)

Over the years, participants in the discussion and analysis have included a broad range of industry organizations. Discussions have been co-chaired by ITS America and the Satellite Industry Association (SIA).[155] Meetings began in May of 2003 and continued through 2006. Independent analyses[156] resulted in the preparation and filing of a proposed spectrum sharing agreement in 2008. Known as the Proposed DSRC/FSS Earth Station Spectrum Sharing Protocol, the agreement proposes how to accomplish in-band sharing between DSRC and FSS operations.

The agreement acknowledges that DSRC and FSS uplink operations are co-primary in 5.9 GHz DSRC band and that there may be potential for interference to DSRC stations from FSS earth stations. To facilitate band sharing, the agreement offers a process for accurate, comprehensive, and up-to-date site information about DSRC and FSS earth stations to be available to licensees. Part 1 of the agreement offers a basis for how the systems will operate; provides appropriate interference criteria; and presents analyses that determine the potential for interference.[157] Part 2 offers a procedural approach to spectrum sharing through definition of the rights and responsibilities of the parties under various operating conditions. These conditions included: (a) new DSRC deployments in the 5.9 GHz DSRC band; (b) new FSS earth station deployments; and (c) modifications to existing FSS earth station deployments. The agreement also proposes rules.

In 2004, a petition for Reconsideration of DSRC Technical and Service Rules was filed and sought an adoption of an "active spectrum management" process to aid in the registration and siting of DSRC stations to avoid potential harmful interference. While the petitioners noted that this process would help mitigate any potential harmful interference from FSS operations in the 5.9 GHz DSRC band, industry discussions were ongoing to determine sound engineering practices for locating DSRC sites to minimize the potential for harmful interference. As a result,

---

[155] Participants for DSRC representation included: ITS America, ARINC, the American Association of State Highway Transportation Officials (AASHTO), Johns Hopkins University/Applied Physics Lab, Mitretek, and Squire Sanders (legal counsel). Participants for FSS representation included: SIA, PanAmSat, Intelsat, SES New Skies, SES Americom, and Steptoe & Johnson (legal counsel). A third party, Comsearch, also participated.

[156] Supporting technical studies included:
- A Johns Hopkins University – Applied Physics Lab, Report on Interference Assessment of In-Band Fixed Satellite Service Earth Station Unlinks to Dedicated Short Range Communications (DSRC) Service Operations (Version 1.0) (2004) (submitted to FCC)
- ARINC, Assessment of Potential Interference to the Fixed Satellite Service (FSS) from the Intelligent Transportation Systems (ITS) Dedicated Short Range Communications (DSRC), Final Report (Version 1.1) (2006) (not provided to FCC).

[157] The determination is based on existing concepts of frequency analysis and coordination for FSS (Part 25) and fixed microwave (Part 101) stations.

the FCC declined to adopt any rules for coordination between FSS and DSRC, pending the results of the industry-to-industry discussions.[158]

In 2010, USDOT met with the FCC to provide ex parte support for the proposed DSRC/FSS Earth Station Spectrum Sharing Protocol. The DSRC industry participants, represented by ARINC, AASHTO and ITS America; and the FSS participants, represented by the SIA, Intelsat, Inmarsat, SES, and Hughes met with the FCC in 2011 for the same reason—to reiterate support for pending proposed DSRC/FSS Earth Station Spectrum Sharing Protocol.

In the Summer/Fall of 2013, the SIA, ARINC, AASHTO, ITS America, and USDOT met to discuss the existing protocol. Discussions regarding these issues are ongoing.

2003-2011: Service Rules

2003: Establishment of the Service Rules

In 2003, the FCC adopted service rules to govern the licensing and use of the 5.850-5.925 GHz band (5.9 GHz band) for the DSRC Service. Based on the USDOT vision of including *"...DSRC units in every new motor vehicle for lifesaving communications,..."*[159] the FCC implemented the following:

- Concluded that it is possible to license both public safety and non-public safety use of the 5.9 GHz DSRC band on all channels, subject to priority for safety/public safety;
- Adopted open eligibility for licensing and technical rules, most of which are embodied in the ASTM-DSRC standard, aimed at creating a framework that ensures priority for public safety communications;
- Adopted a basis for licensing 5.9 GHz DSRC RSUs, communication units that are fixed along the roadside under subpart M (Intelligent Transportation Radio Service) of Part 90 of the Commission's Rules. Licensees will receive non-exclusive geographic-area licenses authorizing operation on seventy megahertz of the 5.9 GHz DSRC band;
- Adopted a framework whereby licensees would register RSUs by site and segment(s);
- Adopted a basis for licensing On-Board Units (OBUs), in-vehicle communications units, by rule under new subpart L of Part 95 of FCC Rules;
- Adopted the standard supported by most commenters and developed under an accredited standard setting process (ASTM E2213-03 or "ASTM-DSRC") to ensure

---

[158] There are more topics covered in the Petitions for Reconsiderations. The full response is in FCC 06-110, located at: https://apps.fcc.gov/edocs_public/attachmatch/FCC-06-110A1.pdf.   (Last accessed June 2015)
[159] FCC Report and Order 2003.

interoperability and robust safety/public safety[160] communications among these DSRC devices nationwide:

- o FCC noted that the ASTM-DSRC Standard included a band plan and technical rules;

- o FCC noted that the band plan reflects a harmonization with Canada and Mexico and that it is divided into channels that are adequate to support the fundamental band communications needs;[161]

- o FCC noted that adopting a standard "…*is appropriate for four reasons:*

  - ▪ *Interoperability;*

  - ▪ *Robust safety/public safety communications and interference management;*

  - ▪ *To promote deployment of DSRC while reducing costs and ensuring that an adequate market develops for equipment; and*

  - ▪ *Consistency with Congressional intent.*"[162]

- o The FCC concluded that compliance with the standard would require compliance with certain technical parameters, such as power limits and receiver performance specifications.

The FCC also concluded that it was *"…premature to adopt rules that reserve certain service channels for specific applications…[that] channel assignments are best addressed under the priority levels of the Control Channel protocol…[to]…give transportation experts additional flexibility in system design… [without]…a negative impact on interoperability."* The FCC further noted that *"…DSRC system design is in its infancy and we expect further development and refinement. Thus, we may need to revisit this issue in the future once we have gained more experience with DSRC operations."*[163]

2006: Amendments to the Service Rules

In 2004, a number of stakeholders filed petitions for reconsideration or clarification of some of the elements of this Rule. Specifically, they asked the FCC to:

---

[160] The terms "safety/public safety" communication are used interchangeably because FCC notes that DSRC Service involves both safety of life communication transmitted from any vehicle, e.g., vehicle-to-vehicle imminent crash warnings, as well as communication transmitted by public safety entities, e.g., infrastructure-to-vehicle intersection collision warnings.
[161] FCC Report and Order, 2003, p.17.
[162] FCC Report and Order 2003, p. 11.
[163] FCC Report and Order 2003, p.18.

- Modify its site registration process to include certain active spectrum management techniques that could identify harmful interference between stations prior to deployment or operation;

- Modify its Universal Licensing System (ULS) to accommodate active registration or consider whether one or more third parties should function as site registration database managers;

- Require DSRC licensees to provide a notice of construction within 12 months after registration, and assign priority rights based on the date of construction notification, rather than on the date of registration in the database;

- Designate Channel 172 exclusively for high-availability, low latency public safety communications, and designate Channel 184 for longer-range, high power public safety DSRC systems;

- Revise the DSRC Class D emission mask, and amend the Rules to create a separate class of OBUs for exclusive use by public safety eligible communications;

- Revise the Rules with regard to an antenna height correction factor requirement intended to minimize potential interference;

- Require that dual-band DSRC devices must be uniquely identified to provide DSRC services in the 5.9 GHz band; and

- Keep Docket WT 01-90 open for future consideration of revisions to the ASTM Standard.

Reconsideration led the FCC to the following decisions:[164]

- **New channel designations**. The FCC designated Channel 172 (frequencies 5.855-5.865 GHz) both for vehicle-to-vehicle collision avoidance or mitigation and other safety of life and property applications; and designated Channel 184 (frequencies 5.915-5.925 GHz) exclusively for high-power, longer-distance communications to be used for public safety applications involving safety of life and property, including road intersection collision mitigation.

- **Site construction and priorities**. FCC amended the rules to require licensees to file a notice of construction with the Commission for each site registered and to clarify that site priority attaches to prior registered sites that have been fully constructed within the requisite 12 month construction period.

- **Increased power**. FCC amended the power reduction rule to only apply to DSRC Roadside Unit antenna height only between eight and 15 meters, thereby providing increased flexibility and reduced implementation costs.

---

[164] Filed respectively by the 3M Company, ARINC Incorporated, Intelligent Transportation Society of America and Johns Hopkins University Applied Physics Laboratory on September 2, 2004.

At that time, the FCC declined to:

- Adopt rules that would implement a software-based prior frequency coordination protocol that directs or recommends that licensees use particular service channels, or that would establish a third party database manager to coordinate and maintain site registrations.

- Amend the current emission mask applicable to DSRC Class D devices, pending further developments and recommendations from the ASTM DSRC Standards Writing Group.

- Adopt rules governing frequency coordination between DSRC licensees and Fixed Satellite Services (FSS) licensees, pending results of studies and ongoing industry discussions; and

- Require dual-band DSRC devices to be uniquely identified in order to use DSRC services in the 5.9 GHz band.

The Commission did, however, reiterate in a number of places that more study is needed and that they retain the discretion to revisit matters.[165]

Further Channel 172 Definition

In their 2006 amendment to the report and order, the FCC offered the following:

"...we agree that there should be an exclusive-use DSRC channel (Channel 172) for public safety applications involving safety of life and property, including vehicle-to-vehicle collision avoidance and mitigation. Were this channel shared and only given priority when needed, the requirement to electronically identify and execute the priority event--even if measured only in milliseconds--could result in an otherwise avoidable vehicular collision. By dedicating Channel 172 for public safety applications, we significantly reduce the potential for interference that would otherwise be expected were the channel shared with non-public safety applications, which in turn reduces the chance that a few milliseconds communications delay could defeat measures crucial to avoiding a collision between vehicles."[166]

In April of 2011, USDOT, CAMP, ITS America, AASHTO, and ARINC provided the FCC with test results that revealed significant errors and gaps in communications when employing a channel

---

[165] These changes are documented in the FCC Memorandum Opinion and Order 06-110 (rel. July 26, 2006). http://apps.fcc.gov/ecfs/document/view;jsessionid=DylJQdGhJvC11xTXyyt2rT3K7Z01GPlp4FtXzB71nBk9lFgnf2CK!-856245186!973241960?id=6518416910. (Last accessed 2015)

[166] Located at: http://apps.fcc.gov/ecfs/document/view;jsessionid=DylJQdGhJvC11xTXyyt2rT3K7Z01GPlp4FtXzB71nBk9lFgnf2CK!-856245186!973241960?id=6518416910, p.9. (Last accessed June 2015)

switching methodology. Tests compared channel switching against employing Channel 172 as a dedicated safety communications channel. The FCC accepted the details presented in this meeting which has allowed stakeholders to develop devices and applications with the appropriate protocols.

D.I.C. 2012-Present DSRC Operations

2012: Notification to FCC of Field Testing in Support of DSRC

In 2012, USDOT met with the FCC to provide an update on efforts relating to DSRC and the development of Connected Vehicle technologies. USDOT conveyed several key points. First, ITS remains a priority for USDOT and its operating administrations, including NHTSA, the Federal Transit Administration (FTA), FHWA, and the Federal Motor Carrier Safety Administration (FMCSA). USDOT noted its investments in ITS research, testing, and other endeavors to help enable a connected transportation environment that will improve safety and mobility. USDOT affirmed that transportation safety is the top priority of the Department, and that V2V and V2I communications have the potential to significantly reduce vehicle crashes and fatalities. In the Department's view, these public benefits support the continued allocation of spectrum for DSRC.

USDOT provided the FCC with further information about its progress in analyzing the benefits of ITS and in testing this technology in real-world scenarios. Among various other endeavors, USDOT described a Connected Vehicle Safety Pilot Model Deployment to begin in late August 2012 in Ann Arbor, Michigan. The test was to involve approximately three thousand vehicles, permitting USDOT to collect data for one year. Finally, USDOT explained that NHTSA has committed to a 2013 agency decision on whether the V2V safety technology (of which DSRC is a foundational element) is sufficiently developed to support rulemaking for light vehicles; with a similar 2014 agency decision point on heavy vehicles.[167]

2013: FCC NPRM on Sharing with Unlicensed Devices

As progress toward Connected Vehicle testing and planning for future implementations was moving forward, on June 28, 2010, President Obama directed the Secretary of Commerce to work with the FCC to identify and make available 500 megahertz of spectrum over the next 10 years for wireless broadband use. On February 22, 2012, the President signed the Middle Class Tax Relief and Job Creation Act of 2012 into law. The Act requires the Assistant Secretary of Commerce (through NTIA), in consultation with the Department of Defense (DoD) and other impacted agencies, to evaluate spectrum-sharing technologies and the risk to Federal users if

---

[167] http://apps.fcc.gov/ecfs/document/view?id=7021995268. (Last accessed June 2015)

Unlicensed-National Information Infrastructure (U-NII) devices were allowed to operate in these bands.

The most common types of U-NII devices include those that use Wi-Fi communication. These devices, in general, operate without a license, but are not supposed to interfere with licensed devices, and have no interference protection. The NTIA was required to issue a report eight months after enactment (October 22, 2012) on the portion of the study on the 5.350-5.470 GHz band. The Act requires the report on the portion of the study on the 5.850-5.925 GHz band no later than 18 months after enactment (August 22, 2013). NTIA published in January 2013 the results of their initial study evaluating known and proposed spectrum-sharing technologies and the risk to Federal users if the FCC allows U-NII devices to operate in the 5.850-5.925 GHz band. Based on a qualitative perspective, the NTIA report identified a number of risks to FCC-authorized stations operating DSRC systems for ITS in the 5.850-5.925 GHz band and suggested mitigation strategies to explore. The NTIA has indicated they are combining both studies into one report.

On April 10, 2013, the FCC published in the Federal Register its NPRM to revise Part 15 of its Rules to permit U-NII devices in additional portions of the 5 GHz spectrum, including the 5.850-5.9250 GHz, so as to "increase wireless broadband access and investment." While the FCC NPRM proposes permitting U-NII devices in the 5.850-5.9250 GHz band, DSRC, as the incumbent, is proposed to retain its primary allocation of the band – U-NII devices would have to operate on a secondary, non-interfering basis. In June 2013, at the request of USDOT, NTIA forwarded to the FCC the comments and concerns that USDOT expressed relating to the deployment and protection of DSRC in the 5.850-5.925 GHz band.[168]

The Institute for Electrical and Electronics Engineers (IEEE) 802 standards committee has established a working group, known as the IEEE 802.11 DSRC Coexistence Tiger Team that provides an international multi-stakeholder technical forum and includes industry experts previously involved in developing standards for both wireless local area networks and vehicular wireless communications. NTIA's January 2013 Report indicated that NTIA will follow up with quantitative studies in connection with domestic and international[169] regulatory proceedings involving the 5350-5470 MHz, 5850-5925 MHz, and other bands. The Tiger Team's meetings have provided a venue for evaluating the coexistence proposals that have been proffered to date. USDOT participates in these calls to help the agency evaluate proposals.

---

[168] This comment can be found on the FCC's docket at: http://apps.fcc.gov/ecfs/document/view?id=7022424618 (last visited July 31, 2015).

[169] From an international perspective, spectrum sharing in the 5.850-5.925 band was removed from discussions by WRC 15 over concerns with interference. World radio communication conferences (WRC) are normally held every three to four years to review, and, if necessary, revise the Radio Regulations, the international treaty governing the use of the radio-frequency spectrum, including services using satellite orbits. NTIA is responsible for coordinating the US Federal government's participation in the International Telecommunication Union's WRC-15.

USDOT will work with the NTIA in a formal manner to review and analyze proposals. Once this analysis is complete, DOT, along with the NTIA and the FCC, will be better positioned to assess how the proposed changes to existing rules and regulations will impact DSRC. At this time, there is insufficient information about how U-NII devices would detect DSRC devices, and how U-NII devices would yield access to the channels within the 5.850-5.925 GHz band. As a result, USDOT cannot insure that the unlicensed devices will not jeopardize safety. If permitted to operate in the 5.9 GHz DSRC band, USDOT will look to the FCC and the NTIA to insure that unlicensed devices do not prevent or delay access to the desired channel, or otherwise pre-empt DSRC safety applications. USDOT will continue to work with the FCC to explore different avenues to facilitate and encourage inter-industry and inter-agency collaborative efforts to assess sharing in the 5.850-5.925 GHz band.

2014: NHTSA Decision on Light Vehicles

On February 3, 2014, NHTSA announced that it will begin taking steps to enable V2V communication technology for light vehicles.[170] NHTSA has worked in close partnership in this research with other USDOT agencies, including the Office of the Assistant Secretary for Research and Technology and the Federal Highway Administration, and with several leading auto manufacturers and academic research institutions, who have invested significant resources into developing and testing V2V technology. The collaboration of government, industry and academia is critical to ensure V2V technology's interoperability across vehicles.

> *"Vehicle-to-vehicle technology represents the next generation of auto safety improvements, building on the life-saving achievements we've already seen with safety belts and air bags. By helping drivers avoid crashes, this technology will play a key role in improving the way people get where they need to go while ensuring that the United States remains the leader in the global automotive industry."*
>
> *—U.S. Transportation Secretary Anthony Foxx*

NHTSA is also considering future actions on crash avoidance safety technologies that rely on on-board sensors. Those technologies are eventually expected to blend with the V2V technology. NHTSA issued an Interim Statement of Policy in 2013[171] explaining its approach to these various streams of innovation. In addition to enhancing safety, these future applications and technologies could help drivers to conserve fuel and save time.

---

[170] http://www.nhtsa.gov/About+NHTSA/Press+Releases/2014/USDOT+to+Move+Forward+with+Vehicle-to-Vehicle+Communication+Technology+for+Light+Vehicles. (Last accessed June 2015)
[171] www.nhtsa.gov/staticfiles/rulemaking/pdf/Automated_Vehicles_Policy.pdf (Last accessed June 2015)

"V2V crash avoidance technology has game-changing potential to significantly reduce the number of crashes, injuries and deaths on our nation's roads. Decades from now, it's likely we'll look back at this time period as one in which the historical arc of transportation safety considerably changed for the better, similar to the introduction of standards for seat belts, airbags, and electronic stability control technology."

—NHTSA Acting Administrator David Friedman

In announcing the agency's intentions, NHTSA noted that V2V technology will not collect or store any data identifying individuals or individual vehicles, nor will it enable the government to do so. There is also no data in the safety messages exchanged by vehicles or collected by the V2V system that could be used by law enforcement or private entities to personally identify a speeding or erratic driver. The system will not enable tracking through space and time of vehicles linked to specific owners or drivers. Third parties attempting to use the system to track a vehicle would find it extremely difficult to do so. The system will not collected any financial information, personal communications, or other information linked to individuals. Finally, the system will not provide a "pipe" into the vehicle for extracting data[172] In fact, the system as contemplated contains several layers of security and privacy protection to ensure that vehicles can rely on messages sent from other vehicles and that a vehicle or group of vehicles would be identifiable through defined procedures only if there is a need to fix a safety problem.

NHTSA finalized its analysis and published a research report on V2V communication technology for public comment in August 2014. The report included results of the Department's research findings in several key areas including technical feasibility, privacy and security, and preliminary estimates on costs and safety benefits. NHTSA is now working on a regulatory proposal that would require V2V devices in new vehicles in a future year, consistent with applicable legal requirements, Executive Orders, and guidance. USDOT believes that the signal this announcement sends to the market will significantly enhance development of this technology and pave the way for market penetration of V2V safety applications.

D.I.D.   Upcoming Policy Milestones

Future: Upcoming Policy Milestones—NHTSA Decision on Heavy Vehicles

Following its decision for light duty vehicles, the agency also intends to make a decision concerning the disposition of V2V technology with regard to heavy vehicles. The heavy vehicle research has been performed in parallel with the light vehicle research. The interoperability,

---

[172] NHTSA: "Vehicle-to-Vehicle Communications: Readiness of V2V Technology for Application, DOT HS 812 014, at: http://www.nhtsa.gov/staticfiles/rulemaking/pdf/V2V/Readiness-of-V2V-Technology-for-Application-812014.pdf

security, and safety application research associated with light vehicles directly supports the heavy vehicle research. Interoperable devices (both integrated and retrofit) have been installed on heavy trucks and buses and tested during the Safety Pilot Model Deployment. Heavy vehicle driver clinics have been conducted to obtain feedback from professional drivers about V2V crash avoidance systems.

Heavy trucks tests have included 19 trucks—three industry Freightliner Cascadia trucks with integrated technologies; and 16 fleet vehicles with retrofit safety devices. Another 150 vehicles were outfitted with vehicle awareness devices. Testing was performed on:

- Forward Collision Warning (FCW) application
- Electronic Emergency Brake Light (EEBL) application
- Curve Speed Warning (CSW) application
- Blind Spot Warning (BSW) application
- Intersection Movement Assist (IMA) application
- Bridge Height Information application

Tests included three University of Michigan buses equipped with retrofit devices that were driven by 61 drivers; and 100 transit buses with VADs to test the following for transit:

- FCW
- EEBL
- CSW
- Pedestrian in Signalized Crosswalk Warning
- Vehicle Turning Right in Front of Bus Warning

Photo Sources: Stock

Figure D-1. Safety Pilot Model Deployment – Heavy Vehicles

These data are being used to support an evaluation of the V2V technology while the agency continues finalizing the research to better understand the operational requirements for these vehicles and how they might differ from light-duty vehicles.[173]

---

[173] NHTSA: "Vehicle-to-Vehicle Communications: Readiness of V2V Technology for Application, DOT HS 812 014, at: http://www.nhtsa.gov/staticfiles/rulemaking/pdf/V2V/Readiness-of-V2V-Technology-for-Application-812014.pdf (last accessed July 2015)

For example, installation and testing under real-world conditions identified unique requirements associated with integrating or retrofitting heavy vehicles with Connected Vehicle technologies, including:

- DSRC device installation was found to vary when retrofitting light versus heavy vehicles. Device mounting locations depend on 1) antenna location; 2) device type; 3) proximity to power tap locations; and 4) vehicle type. For instance, mounting the DSRC antennas on the side mirrors of the buses/HV in some instances caused issues with the vehicle body blocking the DSRC communications.
- Safety applications and messages had to be tuned differently due to vehicle dynamics. For instance, the BSM did not correctly model articulated buses/trucks. The BSM treated vehicles as a rigid rectangular body. However, research is underway to include BSM capability to model articulated vehicles as multiple rectangular bodies.

These results will be taken into consideration as NHTSA makes its determination in 2015.

2016: FHWA Guidance

In 2013, the FHWA announced that the agency would develop Guidance for State and local DOTs to facilitate planning, deployment, operations, and maintenance of new Connected Vehicle technologies and applications. The emphasis of the Guidance is on V2I safety, noting that infrastructure components will also support mobility and environmental benefits. The guidance will include, among other subjects:

- Guidance on use of Federal-Aid funds;
- Discussion of use of a range of applications and guidance on the benefits of applications that can provide "day-one" impacts;
- DSRC RSU licensing and siting guidelines;
- Discussion of backward compatibility and legacy systems;
- Guidance on use of the Connected Vehicle Reference Implementation Architecture and Standards, security, and privacy; and
- Guidelines for transportation planning and project investment.

The first version of the FHWA Guidance is grounded in two important research efforts and their results:

- Safety Pilot Model Deployment results
- AASHTO's Footprint Analysis.

FHWA provided draft Guidance for stakeholder feedback at a public workshop in Fall of 2014. Using the feedback, FHWA is in the process of making revisions to release a version to guide CV Pilot implementers and early adopters. The CV Pilot evaluation results will inform a second revision in 2017.

2014: Safety Pilot Model Deployment Test Results related to RSUs

Twenty-six 5.9 GHz DSRC RSUs were tested as part of the Safety Pilot Model Deployment tests. The 5.9 GHz DSRC RSUs are all technically capable of both storing and forming messages. Testing for RSUs included the following:

- To test how vehicles received the messages to support V2I applications (for instance, for signal phasing and timing (SPaT), curve warnings, and curve speed warnings). Additional applications tested from a V2I perspective included:
    - Signal priority for transit and emergency vehicles;
    - Roadway maintenance;
    - Density of pedestrian traffic; and
    - Traffic signal timing.
- To test the capability of supporting communications between on-board vehicle equipment (OBEs) and a security credential management system (SCMS)[174] as a means of receiving updated security certificates.[175]

The Model Deployment also evaluated the use of existing 3G/4G cellular networks to provide vehicles with updated security certificates, because USDOT wanted to examine the feasibility of supporting communications between vehicles and the SCMS though an existing communications infrastructure. Although a nationwide network of 5.9 GHz DSRC RSU devices does not exist at this time and Congress has yet to allocate funds to build such a network, existing 3G/4G cellular networks could potentially be used to support communications between vehicles and the SCMS if a nationwide network of DSRC RSU devices is not available.[176] That said, there still remains parts of the Nation where cellular coverage is non-existent and citizens in such areas would be excluded from the safety benefits delivered by Connected Vehicle applications.

---

[174] All DSRC RSUs used in the Safety Pilot Model Deployment conformed to "5.9 GHz DSRC Roadside Equipment" Device Specification Version 3.0. *See:* http://www.its.dot.gov/safety_pilot/pdf/T-10001-T2-05_RSE_Device_Design_Specification_v30.pdf (Last accessed June 2015).
[175] The security system used in Safety Pilot Program did not involve distribution of a Credential Revocation List (CRL) but used a "test" CRL to prove transmittal, receipt, and action.
[176] NHTSA: "Vehicle-to-Vehicle Communications: Readiness of V2V Technology for Application, DOT HS 812 014, at: http://www.nhtsa.gov/staticfiles/rulemaking/pdf/V2V/Readiness-of-V2V-Technology-for-Application-812014.pdf. (Last accessed June 2015)

2014: AASHTO Footprint Analysis[177]

As key Federal policy decisions relating to Connected Vehicle safety are moving forward, AAHSTO has partnered with USDOT and Transport Canada to develop a Connected Vehicle Field Infrastructure Footprint Analysis that would provide information to agency investment decision-makers. The analysis helps State and local transportation agencies understand what they will need to prepare for implementation of Connected Vehicle environments, and what investments may need to be made in preparation. AASHTO's work in this analysis has been performed in collaboration with its Connected Vehicle Deployment Coalition; a group comprising representatives from a number of state and local transportation agencies. In addition, the development of Connected Vehicle deployment scenarios engaged a broader community of state and local agency participants.

As the NHTSA decision greatly increases the likelihood of Connected Vehicle environments being implemented, AASHTO's analysis anticipates the emergence of a nationwide base of suitably-equipped vehicles, and V2I applications becoming a practical reality with deployment of a suitable field infrastructure. A Connected Vehicle infrastructure deployment is envisioned to generally include:

- Roadside communications equipment (for DSRC or other wireless services) together with enclosures, mountings, power, and network backhaul;

- Traffic signal controller interfaces for applications that require SPaT data;

- Systems and processes required to support management of security credentials and ensure a trusted network;

- Mapping services that provide highly detailed roadway geometries, signage, and asset locations for the various Connected Vehicle applications;

- Positioning services for resolving vehicle locations to high accuracy and precision; and

- Data servers for collecting and processing data provided by vehicles and for distributing information, advisories, and alerts to users.

Some elements, such as traffic signal interfaces or roadside equipment to send infrastructure information or to receive DSRC messages broadcast from vehicles, are primarily in the domain of State and local DOTs. Other elements of the overall Connected Vehicle system, particularly those necessary for vehicle-based safety applications, may be provided by the automotive industry, and the elements associated with security management could be provided by a third-party entity. These specifics are still evolving.

---

[177] Analysis from AASHTO's *National Connected Vehicle Field Infrastructure Footprint Analysis- Final Report (FHWA-JPO-14-125)*, AASHTO/USDOT; June 27, 2014 at http://stsmo.transportation.org/Documents/AASHTO%20Final%20Report%20 v1.1.pdf (Last accessed June 2015).

AASHTO envisions a mature Connected Vehicle environment by 2040, by which time a large majority of vehicles on the roadway will be connected. Using the historical rate of ITS deployment as a basis and assuming no major changes to transportation policy or funding, the following is reasonable to expect by 2040:

- Up to 80% (250,000) of traffic signal locations could be V2I-enabled;
- 25,000 other roadside locations at non-intersections could be V2I-enabled;
- Accurate, real-time, localized, traveler information will be available on 90% or more of roadways; and
- Next-generation, multimodal, information-driven, active traffic management will be deployed system-wide.

As field deployments will be more viable and effective when they support multiple applications, the analysis identifies the common aspects of the potential applications; such as leveraging the physical infrastructure (e.g., a roadside unit) or the information components (e.g., the basic safety message broadcast by vehicles) that can support a group of applications with common requirements. This approach affects key design and implementation considerations and may affect the cost and complexity of deployment.

Finally, the analysis illustrates that many of the identified non-safety applications could be deployed with either DRSC or cellular communications between the vehicle and infrastructure. Although future research and development may conclusively identify the preferred means of communications for each application, this analysis presumes that most V2I applications, with the exception of time-sensitive safety applications, are viable with either DSRC or cellular.

D.II. DSRC: Technical History

The following section offers a summary of the key technical milestones in developing DSRC for operational use. The appendix is structured, as shown:

<u>DSRC Technical Milestones</u>

Research and Development of DSRC Technologies and Applications

- Requirements
- 2006-2009: DSRC and Autonomous Sensing Safety System Analysis
- 2010-2012 DSRC Technical Issues Related to Interoperability, Scalability, Security, and Data Integrity/Reliability
- Risk Assessments and Design of a Security Credential Management System

Lessons Learned from Testing

- 2008:Proof of Concept
- 2012-2013 Model Deployment

Maturing the Technologies

- Completing Performance Requirements
- Standards
- Production of Technologies

Technical Milestones

- 2014: Scalability Test Results
- 2014: ITS World Congress Testing

D.II.A   Research and Development of DSRC Technologies and Applications

For over a decade, USDOT has worked in partnership with CAMP on vehicle safety communications. Various members came together to form the vehicle safety consortium (VSC)[178] and conducted three rounds of research focused on the development of:

- Vehicle safety communications applications;
- Establishment of testing procedures and standards;
- Testing of the interoperability, reliability, and accuracy of equipment and DSRC-based vehicle safety communications systems; and
- Design of the deployment model.

The following sections provide an overview of major technical milestones that the VSC Consortium has achieved in developing and testing DSRC technology.

D.II.A.i.   2005-2006: DSRC Enabled Applications Communications Requirements

In 2005, the first VSC (VSC 1) investigated potential applications and relevant communications requirements for DSRC-based communications. VSC 1 first compiled a comprehensive list of safety applications that members believed would benefit from or would be enabled by the availability of DSRC, grouping safety application scenarios into categories:

- Intersection Collision Avoidance;
- Public Safety, Sign Extension;
- Vehicle Diagnostics and Maintenance; and
- Information from Other Vehicles.

The Consortium also identified applications in non-safety categories. The Consortium ranked this initial list of more than 75 applications scenarios based on potential safety benefits, estimated deployment timeframe, estimated market penetration, and estimated cooperation from infrastructure and/or other vehicles.[179]

---

[178] Members of the VSC Consortium have included BMW of North America, LLC, DaimlerChrysler Research and Technology North America, Inc., Ford Motor Company, General Motors Corporation, Honda Research and Development Americas, Inc., Hyundai-Kia America Technical Center, Inc., Mercedes-Benz Research and Development North America, Inc. Nissan Technical Center North America, Inc., Toyota Technical Center USA, Inc., and Volkswagen of America, Inc.

[179] For its potential safety benefits analysis, the Consortium chose to rely upon the Functional Years Lost metric, defined by General Motors in its 44 Crashes report as "the number of years lost to fatal injury plus the number of years of functional capacity lost to nonfatal injury: General Motors (1997). *44 Crashes, v.3.0.* Warren, MI: NAO Engineering, Safety & Restraints Center, Crash Avoidance Department. In applying this metric, the Consortium considered both the benefit opportunity (the safety benefits of deploying DSRC-enabled safety applications on all new

The Consortium documented detailed communications requirements for the eight application scenarios with the highest estimated benefits. In general, it found that the most promising safety applications required a maximum 100 millisecond latency period and the ability to send data packets of up to 430 bytes a distance of up to 1,000 meters. It also concluded that the ability to broadcast messages, rather than send point-to-point messages, would be important to ensure that vehicles can easily communicate with other vehicles in close proximity.

As previously mentioned in this report, USDOT has analyzed communications technologies multiple times over the years, based on the understanding that the capabilities are evolving. The first analysis was in 2005 when the Consortium compared 13 wireless communication technologies against the general application requirements and concluded that DSRC was unique in meeting the basic communications requirements for most of the vehicle safety applications. None of the other technologies evaluated came close to meeting all of the safety application requirements; the Consortium identified latency as one of DSRC's most significant advantages over competing technologies.[180]

D.II.A.ii.   2006-2009: DSRC and Autonomous Sensing Safety System Analysis

Between 2006 and 2009, a second round of research built upon the results of the VSC 1 research to evaluate the potential of DSRC-based vehicle safety communications to enhance the capabilities of autonomous safety systems (e.g., radar-based rear-end and road departure collision warning systems) and enable new safety applications. Autonomous safety systems have shown measurable benefits in reducing crashes in field operational tests, but could be far more effective if combined with vehicle positioning and wireless vehicle-to-vehicle communications. The VSC 2 Consortium identified seven crash scenarios that could be addressed by DSRC-based safety systems and assessed the ability of DSRC, in combination with positioning sensors and forward-looking radar, to enhance the capabilities of positioning and forward-looking radar systems operating in isolation.

The VSC 2 Consortium indeed found that DSRC-based communications can overcome limitations associated with autonomous sensing safety systems like forward-looking radar systems. The Consortium identified several limitations of autonomous sensing safety systems, including late confirmation of stopped vehicles, occasional incorrect target detection, and late or false alerts for lead vehicle movements.

As part of this research effort, each Consortium member developed a vehicle test bed that, through testing, was shown to be interoperable with the test beds of other members. The

---

vehicles) and estimated benefits (the safety benefits based on estimates of market penetration). The Consortium also considered the direct costs saved and number of vehicles saved in each application scenario.
[180] http://www.its.dot.gov/research_docs/pdf/59vehicle-safety.pdf, Page 46. (Last accessed June 2015)

Consortium also developed objective test procedures to demonstrate the performance of test beds against minimum performance specifications.

The VSC 2 Consortium made two particularly significant contributions to the development of DSRC:

- First, they established the BSM standard, now defined in the SAE J2735 Message Set Dictionary standard, and made significant contributions to the IEEE 1609.2 (Security Services), IEEE 1609.3 (Networking Services), IEEE 1609.4 (Multi-Channel Operation), and IEEE 802.11p (Wireless Access in Vehicular Environments) standards. The proposed BSM standard contains information about vehicle latitude and longitude, elevation, and positional accuracy, and was specifically designed to convey in a single message all information necessary to support the vehicle safety communication applications identified by VSC 1.

- Second, they began the evaluation of the scalability of DSRC radios, testing several channel configuration methods for operating up to 60 DSRC radio units at one time in close proximity. The Consortium tested four channel configurations—three channel switching schemes based on IEEE 1609.4 and one scheme relying on a single channel dedicated to safety messages—and concluded that the use of a full-time channel dedicated to safety provided superior performance to any of the channel-switching configurations. These tests also found that reducing message packet size and transmission rates lowered error rates.[181] The third round of VSC research built on these results.

D.II.A.iii.   2010-2012: DSRC Technical Issues Related to Interoperability, Scalability, Security, and Data Integrity/Reliability

Between 2010 and 2012, VSC 3 conducted research to address the technical issues related to interoperability, scalability, security, and data integrity of DSRC. The Consortium developed and evaluated methods of achieving interoperability of DSRC equipment by establishing and issuing requirements to four hardware suppliers for on-board equipment (OBE). These requirements related mainly to compatibility with communications standards and vehicle hardware, as well as hardware specifications for OBE components, including the GPS receiver and antenna and DSRC radio and antenna. The Consortium established test cases to ensure conformance of software and OBE hardware to minimum interoperability requirements. In general, the Consortium found that the published DSRC standards were mature and the interpretation of

---

[181]

http://www.nhtsa.gov/DOT/NHTSA/NVS/Crash%2520Avoidance/Technical%2520Publications/2011/811492A.pdf&rct=j&frm=1&q=&esrc=s&sa=U&ei=mpGPU5rtHKfLsASKo4CgDQ&ved=0CBcQFjAA&sig2=JIUXOtUsZUYe9QPqsa4zxw&usg=AFQjCNGOeMPqqa0DviyhEdnYsh6kvaegUw (Last accessed June 2015)

---

these standards by suppliers allowed for interoperability among OBEs provided by different suppliers.

Given the intended deployment of DSRC on the Nation's roadways, many of which experience high volumes in concentrated areas, the Consortium undertook additional scalability testing to evaluate how the technology performed in both congested and uncongested environments. As part of its scalability testing, the Consortium developed two algorithms for controlling the transmission of safety messages. It then tested the performance of these two algorithms against a baseline in driving scenarios that emulated real-world driving situations at congestion levels of 50, 100, and 200 vehicles.

As part of this scalability testing, the Consortium also developed and tested a prototype security system. The system was developed for a number of purposes, including:

- Identifying options for the creation, distribution, and installation of security certificates;

- Addressing the potential for malfunctioning equipment or users with ill-intentions to transmit spurious messages; and

- Determining how to protect privacy within a certificate-based security system.

The basic prototype security system was designed with the objectives of ensuring privacy, achieving authentication through message verification rather than hardware authentication, guaranteeing system robustness and rapid recovery, and then providing the ability to prove the integrity and origin of data in order to report bad actors for certificate revocation. Through the development of this security system prototype, the Consortium also ensured that security would not negatively impact safety, and that it could be scaled up to handle larger volumes of messages. It was also designed to be transparently updated from the user's perspective.

Finally, the Consortium evaluated accuracy and reliability requirements needed for DSRC-based safety applications to function properly. Based on performance specifications and objective test procedures defined by the VSC 2 Consortium, VSC 3 established upper bounds for system performance. The Consortium simulated the propagation of errors through several of the safety application systems identified by VSC 1. The Consortium incorporated these accuracy requirements into SAE J2945 Minimum Performance Requirements, the companion document to the SAE J2735 standard.[182]

---

[182] Ahmed-Zaid, F. et al. "Interoperability Issues of Vehicle-to-Vehicle Based Safety Systems Project (V2V-Interoperability): Final Report." CAMP Vehicle Safety Communications 3. Submitted to the Intelligent Transportation Systems (ITS) Joint Program Office (JPO) and the National Highway Traffic Safety Administration (NHTSA) in response to Cooperative Agreement Number DTNH22-05-H01277. September 30, 2012.

D.II.A.iv. Risk Assessments and Design of a Security Credential Management System

In 2012, a team of OEMs, USDOT personnel, automotive suppliers, and security experts defined a deployment model for a DSRC-based V2V environment. The deployment model consisted of a security credential management system (SCMS), which issued the certificates that identify valid DSRC messages, hardware and connectivity requirements, and provisions for a graceful evolution from initial to final deployment. The team specifically assessed hardware requirements to ensure that:

- The necessary equipment would be both robust from a security perspective and affordable;
- The equipment would be technically viable on a large scale (i.e., 350 million vehicles in the US fleet);
- The privacy of all users would be preserved; and
- There would be efficient misbehavior detection, revocation, and distribution of revocation information to each user.

As part of this work, a risk analysis was performed to identify vulnerabilities and the system's potential exposure to attacks, malicious or otherwise. The team considered the likelihood and impact of these potential attacks from a range of attackers, including talented engineers or cryptographers without detailed inside knowledge, knowledgeable insiders, funded government and non-government organizations, SCMS insiders, unintentional attackers (i.e., an attack due to equipment malfunction), and combinations of these types of attackers. In assessing the likelihood of attack, the team considered the motivations and capabilities of the attacker, the presence of preventative controls, the ease and repeatability of attack, and the availability of emergency backup systems.

These results have formed the basis for the organizational, physical, operational, and policy design of the current SCMS that is under development as a fuller-scale prototype today.[183]

D.II.B. Lessons Learned from Testing

D.II.B.i. 2008 Proof of Concept

A critical result of the previous ITS research (2005-2009) was the establishment of a test environment in Novi, Michigan to conduct a set of proof-of- concept tests using field (roadside)

---

[183] CAMP Vehicle Safety Communications 3. "V2V-Communications Security Project: Risk Assessment and Technical Analysis Report." Submitted to the Intelligent Transportation Systems (ITS) Joint Program Office (JPO) of the Research and Innovative Technology Administration (RITA) and the National Highway Traffic Safety Administration (NHTSA) in response to Cooperative Agreement Number DTFH61-01-X-00014. August 16, 2012.

installations in Michigan and California and a network control center located in a Herndon, Virginia facility. Those test results were instrumental in forming the foundation of this current research agenda, and its recommendations are reflected in the structure of the technology, applications, and policy research programs that support a connected transportation environment.

USDOT and Michigan Department of Transportation jointly developed the Novi test bed in 2008.[184] The site was created to serve as a general test bed and to:

- Evaluate data sampling errors
- Generate data using snapshots
- Study the effect of privacy safeguards on probe data
- Provide information to affected subsets of vehicles
- Broadcast signal phasing and timing information to vehicle approaching intersections
- Design data routing protocols
- Design car-following algorithms less sensitive to information delays
- Provision to USDOT a sample dataset of vehicle snapshots capturing the information that could be collected in a network under full market penetration scenario.

The design of the Michigan test bed was influential in the installation of the Florida affiliated interoperable test bed in Orlando, Florida in 2011. To enhance the utility of the site, the Novi test bed also participated as the testing area for the SafeTrip-21 program. The program aimed to confirm the capability of VII network to provide end-to-end, interoperable communications.[185] SafeTrip-21 utilized the 42 square miles of testing area including 56 DSRC RSUs and 75 lane-miles or arterials and freeways.

Upon completion, the program successfully developed:

- The first vehicle-infrastructure integrated prototype system in a real-world environment.
- The first integrated vehicle-infrastructure architecture and concept of operations that:
  - Described a variety of State and local use case scenarios
  - Identified the DSRC standards that enable system interoperability and facilitate integration

---

[184] http://ntl.bts.gov/lib/51000/51000/51002/DF3F1F7.pdf page 21
[185] ITS Research Results: ITS Program Plan 2008, page 53 (located at: http://ntl.bts.gov/lib/30000/30800/30867/ITS_Research_Results_-_ITS_Program_Plan_2008_-_ITS_Report.pdf). (Last accessed June 2015)

- o Employed the first technologies used in data gathering, transmission and broadcast,
- o Identified a process to migrate the architecture to an open platform.
- A novel approach to large-scale network security that incorporates privacy controls.[186]

As of 2012, the Novi test bed has been updated with equipment to meet current technology standards. The upgrade also allowed users to test applications using close or dispersed communications.[187]

D.II.B.ii.　2012-2013 Model Deployment

The vision of the Safety Pilot Model Deployment was to test V2V safety applications in real-world driving scenarios to support estimation of their effectiveness at reducing crashes, and to ensure that the devices are safe and do not unnecessarily distract motorists or cause unintended consequences. The Safety Pilot also evaluated everyday drivers' reactions, both in a controlled environment through driver clinics, and on actual roadways with other vehicles through the real-world model deployment.

The goals of the Safety Pilot were to support the NHTSA agency decision by obtaining empirical data on user acceptance and system effectiveness; demonstrate real-world Connected Vehicle applications in a data-rich environment; establish a real-world operating environment for additional safety, mobility, and environmental applications development; archive data for additional research purposes; and identify prototype system characteristics that can be improved or that need to be corrected.

The two fundamental components of the Safety Pilot are:

- **Safety Pilot Driver Clinics:** Driver clinics were conducted at six sites across the United States to assess user acceptance of the V2V technology. At each driver clinic, approximately 100 drivers tested in-vehicle wireless technology in a controlled environment, such as a race track. The goal was to determine how motorists responded to and benefitted from in-vehicle alerts and warnings. The driver clinics were conducted from August 2011 through January 2012.

- **Safety Pilot Model Deployment:** The model deployment was conducted in the city of Ann Arbor, Michigan from August 2012 to February 2014. Sponsored by USDOT and conducted by the University of Michigan Transportation Research Institute (UMTRI), the experiment was designed to support estimation of the effectiveness of V2V technology at reducing crashes. Approximately 2,800 vehicles—a mix of cars, trucks, and transit vehicles

---

[186] Ibid, page 45.
[187] http://www.its.dot.gov/newsletter/august2012.htm (Last accessed June 2015)

---

operating on public streets within a highly concentrated area—were equipped with integrated in-vehicle safety systems, aftermarket safety devices, or vehicle awareness devices, all using DSRC to emit wireless signals of vehicle position and heading information. Vehicles equipped with integrated in-vehicle or aftermarket safety devices had the additional design functionality of being able to warn drivers of an impending crash situation involving another equipped vehicle. The graphic on the following page provides an illustration of the types of devices and vehicles deployed during Safety Pilot.[188] V2V devices installed in light vehicles as part of the Connected Vehicle Safety Pilot Model Deployment were able to transmit and receive messages from one another, with a security management system providing trusted and secure communications among the vehicles during the Model Deployment. This was accomplished with relatively few problems given the magnitude of this first-of-its-kind project. The V2V devices tested in the Model Deployment were originally developed based on existing communication protocols found in voluntary consensus standards from SAE and IEEE. NHTSA and others participating in the Model Deployment (e.g., its research partners and devices suppliers) found that the standards did not contain enough detail and left room for interpretation. Therefore, they developed additional protocols which enabled "interoperability" between devices participating in the study. The valuable interoperability information learned during the execution of Model Deployment is planned to be included in future versions of voluntary consensus standards that would support a larger, wide-spread technology rollout.[189]

---

[188] From http://www.drive-c2x.eu/tl_files/publications/3rd%20Test%20Site%20Event%20TSS/4_7%20DRIVE%20C2X%203rd%20Test%20site%20event_Kate%20Hartman_Testing_Users_20130613.pdf (Last accessed June 2015)

[189] NHTSA: "Vehicle-to-Vehicle Communications: Readiness of V2V Technology for Application, DOT HS 812 014, pg. 15, at: http://www.nhtsa.gov/staticfiles/rulemaking/pdf/V2V/Readiness-of-V2V-Technology-for-Application-812014.pdf (Last accessed June 2015)

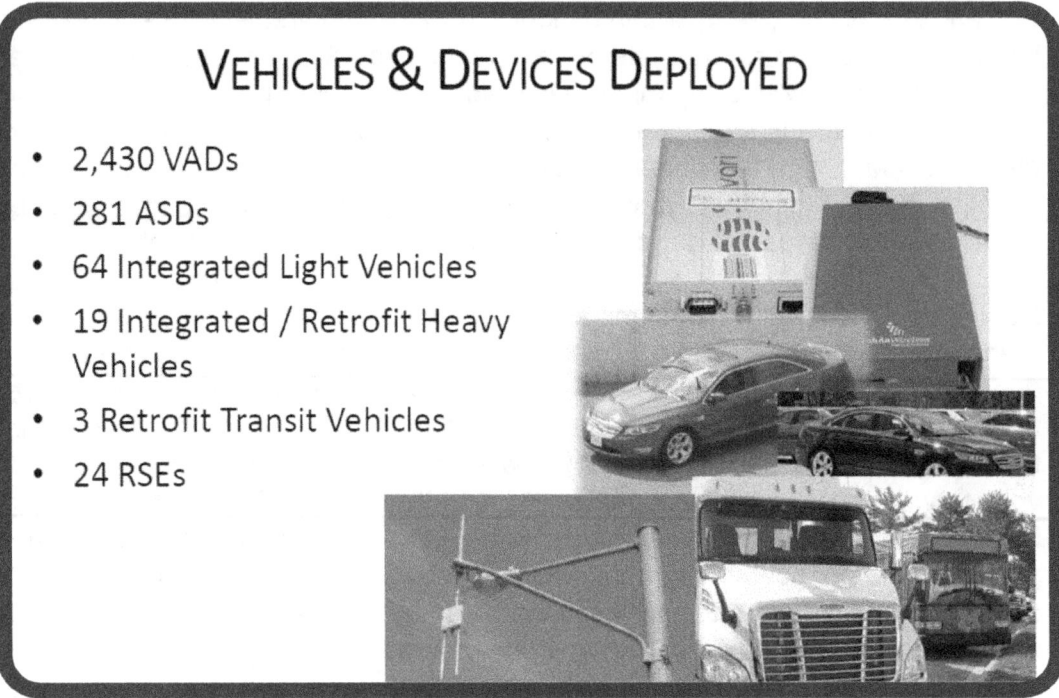

VEHICLES & DEVICES DEPLOYED

- 2,430 VADs
- 281 ASDs
- 64 Integrated Light Vehicles
- 19 Integrated / Retrofit Heavy Vehicles
- 3 Retrofit Transit Vehicles
- 24 RSEs

Figure D-2. Illustration of the Vehicles & Devices in Safety Pilot Photos: USDOT

D.II.C.   Maturing the Technologies

Mature technologies ready for technology transfer to the market place have stable performance requirements with objective test procedures, are supported by stable standards, and have tested prototypes that are production ready. This section offers insight into the status of the Connected Vehicle technologies.

D.II.C.i. Completing Performance Requirements[190]

NHTSA has certain performance measures already available and is also currently working to develop a comprehensive list of DSRC use cases as a basis for developing performance measures.[191] Both CAMP and the Booz Allen Hamilton research have provided important inputs into describing performance requirements.

At its most basic, DSRC must be capable, among other things, of sending and receiving BSMs to/from other vehicles and to infrastructure; of *not* sending or receiving certain types of

---

[190] Summarized from NHTSA: "Vehicle-to-Vehicle Communications: readiness of V2V Technology for Application, DOT HS 812 014, pages 74-75, .at: http://www.nhtsa.gov/staticfiles/rulemaking/pdf/V2V/Readiness-of-V2V-Technology-for-Application-812014.pdf (last accessed June 2015)

[191] Ibid. See: Section IV for more information.

information that might be harmful to the vehicle or to the V2V system (including BSMs, if the system recognizes or the DSRC recognizes itself to be somehow compromised); and of receiving new certificates and software updates. Each of those tasks, in turn, has sub-tasks in order to ensure effective performance. For example, when a DSRC unit sends out a BSM, the BSM needs to:

- Contain the relevant elements and describe them accurately (e.g., vehicle speed; position[192]; vehicle heading; DSRC message ID; etc.);

- Be received quickly enough for the receiving DSRC unit to interpret the message and respond accordingly by triggering safety applications or not; and

- Contain information to indicate that it should be trusted by the receiving DSRC unit and that the message has not altered (e.g., a signed security certificate that is up-to-date).

SAE J2945.1, developed in large part through USDOT funds, contains minimum performance requirements for BSM communication, but not yet for message accuracy, test procedures, or how the data and message would be used (such as message transmission rate or optional data usage in various situations); nor is it certain that they will do so in the future.[193]

D.II.C.ii.   Standards[194]

The following provides an update on the critical standards that enable an interoperable V2V and V2I Connected Vehicle environment:

---

[192] It should be noted that with regard to position, this data comes from GPS. Due to the concerns with jamming, the vehicle will employ plausibility checks from other techniques such as inertial analysis; the safety applications will also employ plausibility checks to ensure that GPS data from other vehicles reflects reality before it uses the data to inform a warning or alert for the driver.

[193] Please see: System Requirements Description, 5.9 GHz DSRC Vehicle Awareness Device Specification, Version 3.6 (Jan. 25, 2012) *at:* http://www.its.dot.gov/newsletter/VAD%20Specs.pdf (Last accessed June 2015) and System Requirements Description, 5.9 GHz DSRC Vehicle Awareness Device Specification, Version 3.6 (Dec.26, 2011) *at* http://www.its.dot.gov/meetings/pdf/T2-05_ASD_Device_Design_Specification_20120109.pdf (Last accessed June 2015)

[194] Summarized from NHTSA: "Vehicle-to-Vehicle Communications: readiness of V2V Technology for Application, DOT HS 812 014, pgs. 99-104 at: http://www.nhtsa.gov/staticfiles/rulemaking/pdf/V2V/Readiness-of-V2V-Technology-for-Application-812014.pdf (Last accessed June 2015)

- **SAE J2735 – DSRC Message Set Dictionary:** The J2735 standard specifics a message set, and its data frames and data elements specifically for use by applications intended to utilize the 5.9 GHz DSRC for Wireless Access in Vehicular Environments (WAVE) communications systems. The data frames and elements have been designed, to the extent possible, to also be of potential use for applications that may be deployed in conjunction with other wireless communications technologies. In 2015, the second published version was released and is usable for deployment.

- **SAE J2945 - DSRC Minimum Performance Requirements:** The J2945.1 standard is part of a future family of J2945.x standards.[195] The current draft standard consists of multiple sections with each section describing the specific requirements for using the BSM for V2V safety applications. A final draft version of J2945.1 exists which includes the minimum communication performance requirements for the Basic Safety Message (BSM). It is anticipated the published version of J2945.1 will be available in the first quarter of 2016.

---

[195] Each J2945.x standard will provide the critical interface information needed to support one or more applications. Associated design specifications for data frames and data elements for the respective J2945.x standards are defined in the SAE J2735-2009 (DSRC Message Set Dictionary standard) and will also be included in future published versions of J2735.

- **IEEE 1609 - Standard for Wireless Access in Vehicular Environments (WAVE):**
  - **IEEE 1609.0—2013 IEEE Guide for Wireless Access in Vehicular Environments (WAVE) Architecture:** The guide was published 11 December 2013.
  - **IEEE 1609.2—2013 Security Services for Applications and Management Messages:** This standard was published 26 April 2013.
  - **IEEE 1609.3—2010 Networking Services:** This standard was published 30 December 2010.
  - **IEEE 1609.3—2010/Cor. 1-2012 Corrigendum 1.** This corrigendum (a separate document) was published 13 July 2012. It corrects errors found in IEEE 1609.3-2010.
  - **IEEE 1609.3—2010/Cor. 2-2014 Corrigendum 2.** This corrigendum (a separate document) was published 23 December 2014. It corrects errors found in IEEE 1609.3-2010.
  - **IEEE 1609.4—2010 Multi-Channel Operations:** This standard was published 7 February 2011.
  - **IEEE 1609.4 Cor. 1 Corrigendum 1.** This draft corrigendum was published in 2014. It corrects errors found in IEEE 1609.4-2010.
  - **IEEE 1609.1 —2012 Identifier Allocations:** This standard was published 21 September 2012.

- **IEEE 802.11-2012 - IEEE Standard for Information technology— Telecommunications and information exchange between systems- Local and metropolitan area networks—Specific requirements Part 11: Wireless LAN Medium Access Control (MAC) and Physical Layer (PHY) Specifications:** This standard was published 29 March 2012.[196]

D.II.C.iii.   Production of Technologies

The Safety Pilot Model Deployment hardware consists of pre-production, prototype components. Many components being used in the Model Deployment could be used to further develop products for full-scale production. In some cases, prototype components used in the Safety Pilot have the appearance and packaging of what could be a regular production device. NHTSA's current understanding, based on discussions with industry OEMs and suppliers, is that securing and preparing manufacturing facilities is the major factor to transitioning from building

---

[196] The previous standard, the IEEE 802.11p-2012 - Medium Access Control and Physical Layer Specifications for WAVE, no longer exists because it was incorporated in a revision of IEEE 802.11. (amendment was actually IEEE 802.11p-2010, not 2012).

prototype components to ramping up to produce mass market components, and that the device in its current form is nearly production-feasible today.[197]

D.II.D. Technical Milestones

D.II.D.i. 2014: Scalability Test Results

Testing of the scalability of the communications network has been conducted under two main projects, the Vehicle Safety Communications- Applications project (VSC-A)[198] and the V2V-Interoperability (V2V-I) project.[199] During VSC-A, 60 vehicles were tested for scalability of the network to see the effects of different data rates, multiple radios, and broadcast frequencies. The V2V-I project tested a grouping of 50, 100, 150, and 200 vehicles under a number of different V2V safety applications in multiple testing locations across the country.

Also tested during the V2V-I project were two algorithms for congestion mitigation.[200] These algorithms are designed to limit the frequency of BSMs broadcast during periods of high channel usage and at the same time ensure that vehicles were able to receive sufficient data to support the safety applications.[201]

Also developed under the V2V-I project was a proof of concept simulator designed to numerically simulate large vehicle networks. The V2V-I project found that even during the 200 vehicle test, at the maximum normal transmit rate of 10 Hz, the channel was not saturated, and all safety applications tested functioned normally. Although channel saturation was not reached, both congestion mitigation algorithms were able to demonstrate decreasing channel congestion while showing good safety application performance.[202]

---

[197] Summarized from NHTSA: "Vehicle-to-Vehicle Communications: readiness of V2V Technology for Application, DOT HS 812 014, pgs. 65-86 at: http://www.nhtsa.gov/staticfiles/rulemaking/pdf/V2V/Readiness-of-V2V-Technology-for-Application-812014.pdf (Last accessed June 2015)

[198] *See:* VSC-A Project Appendix Volumes 1 and 2 for full system requirements and further information. *See also:* Vehicle Safety Communications – Applications (VSC-A), Second Annual Report, January 1, 2008 through December 31, 2008 (DOT HS 811 466) *at:*
http://www.nhtsa.gov/Research/Crash+Avoidance/Office+of+Crash+Avoidance+Research+Technical+Publications
(Last accessed June 2015).

[199] More information can be found in: *Interoperability Issues of Vehicle-to-Vehicle Based Safety Systems Project (V2V-Interoperability) Final Report.*

[200] Algorithm X is a transmission control protocol for scalable V2V safety communications that supports adaptive control of the message transmission rate and transmission power. Algorithm Y controls message transmission rate based on reported CBP from the neighboring vehicles and that measured by the host vehicle. The algorithm adapts the message rate up and down in order to maintain a desired level of channel utilization. For more information, see: *V2V-I Final Report Section 4.2* and *Appendix A, V2V Safety Communications Scalability Algorithms Details.*

[201] V2V-I Final Report, at 79.

[202] V2V-I Final Report, at 79.

Current research has shown that the V2V safety applications perform reliably in test scenarios with up to 200 vehicles in communication range. However, research has yet to estimate fully the number of other DSRC-equipped vehicles that a single DSRC radio would be exposed to in an environment (such as heavy freeway traffic) where channel congestion would be significant. NHTSA is performing additional research on this subject to address that need.[203]

D.II.D.ii.   2014: ITS World Congress Testing

The Southeast Michigan 2014 project test bed gives users the capability to test safety, mobility, and environmental applications, services, and components in an environment using the latest technology standards and architecture consistent with USDOT's V2V and V2I research program. Located in Oakland County, Michigan, the site is equipped with roadside equipment to provide the capability for private sector firms to test applications. The site also provides continuous real-time connectivity among users. It currently covers 45 miles (194 km) comprising 75 linear miles (121 km) of roadways. An expansion will add 6 miles (10 km) of roadways.[204]

The Southeast Michigan 2014 project is advancing the communication and data fundamentals of a Connected Vehicle environment. For the first time, all of the elements are operational— components in vehicles, at the side of the road, and in the back office—to demonstrate information flows that are connected with common security processes. [205]

Importantly, and in response to Congress's question regarding the preferencing of any particular frequency for V2I operations, the information flows are being tested and are "operating medium independent". Cellular wide areas networks, cellular local area networks, DSRC, and satellite media are participating in demonstrating V2I capabilities.

The tests, to date, show that:

- When providing a base level of data for the benefit of all vehicles and travelers that is accessible to all content providers:
    - o Viable media include satellite, WAN, or DSRC in broadcast mode or through internet protocol transactions or streams.
- When providing specific data for private purposes for the benefit of one individual or node within the system:
    - o Viable media include wide area networks (WANs), DSRC or local area networks (LAN); and

---

[203] NHTSA: "Vehicle-to-Vehicle Communications: readiness of V2V Technology for Application, DOT HS 812 014, page 125 at: http://www.nhtsa.gov/staticfiles/rulemaking/pdf/V2V/Readiness-of-V2V-Technology-for-Application-812014.pdf

[204] see more at: http://www.its.dot.gov/testbed/testbed_SEmichigan.htm#sthash.VuoZ3gN5.dpuf

[205] Walt Fehr presentation, April 2014.

- All media can be used in broadcast mode or through internet protocol transactions or streams with access control.

The tests are significant because they are based on a common data exchange process that illustrates the flow and evolution of the data in a "full round trip"—from generation to exchange and fusion with new sources to being provided back to the source device in an enhanced format. Utilizing DSRC on all seven channels and engaging transportation, public safety, and energy applications, are key aspects of the tests that were displayed at the 2014 ITS World Congress. In addition, new tools are resulting; tools that will support and are supporting near-term deployment planning.

Appendix E Definition of Connected Vehicle Enabling Technologies

E.I. Overview of the Enabling Technologies

USDOT and the automobile industry efforts to develop Connected Vehicle technologies have focused on both in-vehicle components as well as external, infrastructure-side components. These components include a security system that manages V2V and V2I communications and ensures trust in the data being transmitted among vehicles and between vehicles and roadside infrastructure. A number of standards have also been developed for how these components function and interact. This appendix provides definitions for the internal and external components as well as the cooperative system standards.

E.I.A. In-Vehicle Components

Examples of the in-vehicle components include:

- **Hardware:** DSRC radios, cables, and antennae to gain wireless connectivty and support data exchange with other vehicles and V2I safety roadside infrastructure, and GPS chips used to determine vehicle location and time. Other hardware includes the secure memory and microprocessor to store security credentials and perform the applications processing.

- **Software:** Software applications that receive, process, and analyze timestamped data broadcasted by other vehciles, such as GPS location, speed, heading and brake status and, based on that analysis, predict when collisions are imminent. Software also includes applications that analyze and process data communicated from V2I safety roadside infrastructure.

- **Driver-Vehicle Interface (DVI):** A DVI that—based on the data analysis conducted by the V2V and V2I software applications—provides a warning to the driver through vehicle-based features such as sounds, lights, or seat vibrations when a collision may be imminent.[206]

---

[206] For example, for potential collisions involving vehicles in a driver's blind spot, one automobile manufacturer's vehicles provided three short low-pitched beeps repeated three times, an orange light in the side view mirror, and a vibration on the side of the driver's seat in the direction of the potential collision.

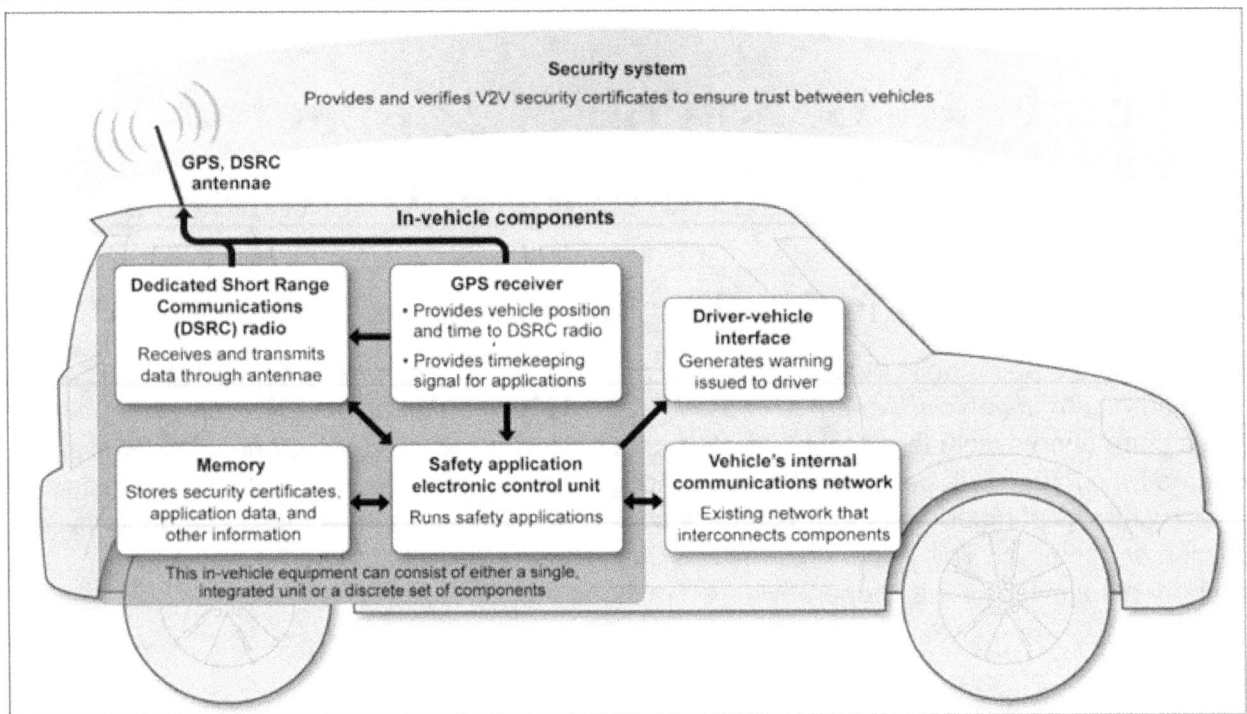

Sources: Crash Avoidance Metrics Partnership and GAO.

Figure E-1. Illustration of DSRC Vehicle-Based Technologies

DSRC devices can be manufactured with a range of capabilities. The text box on the following page describes the range of devices and their functionality, including:

- Integrated on-board equipment (OBEs);
- Retrofit equipment;
- Aftermarket Safety Devices (ASDs); and
- Vehicle Awareness Devices (VADs)

Range of DSRC-based Vehicle Communications Devices

OEM Device: An OEM device is an electronic device built or integrated into a vehicle during vehicle production. An integrated V2V system is connected to proprietary data busses and can provide highly accurate information using in-vehicle information to generate the Basic Safety Message. The integrated system both broadcasts and receives BSMs. If equipped with crash avoidance safety applications, the system can process content of received messages to trigger advisories or warnings to the driver of the vehicle in which it is installed. Because the device is fully integrated into the vehicle at the time of manufacture, vehicles with Integrated Safety Systems could potentially provide haptic warnings to alert the driver (such as tightening the seat belt or vibrating the driver's seat) in addition to audio and visual warnings provided by the aftermarket safety devices that can be integrated into the visual displays and audio systems. It is expected that the equipment required for an integrated OEM V2V system would consist of a general purpose processor and associated memory, a radio transmitter and transceiver, antennas, interfaces to the vehicle's sensors, and a GPS receiver. Such integrated systems are capable of being reasonably combined with other vehicle-resident crash avoidance systems to exploit enhance the functionality of both types of systems.

Retrofit Devices (RSDs): RSDs are defined as those devices that are integrated into the vehicle after manufacturing. Their level of integration is typically less than an OEM device but more than aftermarket devices. RSDs connect to the vehicle and can receive information from the vehicle's data bus to support operations of various applications on the devices. RSDs were tested with heavy vehicles during the Safety Pilot Model Deployment. The advantage to RSDs, as compared to the other types of aftermarket devices, is that they can potentially perform different or enhanced safety applications or execute more sophisticated applications because they can access a richer set of data (i.e., data from the data bus). For example, having information on the turn signal status from the vehicle provides the device and application an indication of possible driver intent to make a turn, which can help inform the LTA, DNPW, BSW/LCW safety applications Under certain circumstances, a RSD may also be able to utilize existing safety oriented DVIs. RSDs are likely to be dealership installed devices. Therefore, the RSD is considered to be the closest of all of the aftermarket devices to a V2V device integrated into a new vehicle.

Aftermarket Safety Devices (ASDs): Automotive aftermarket devices can be defined as any product added to a motor vehicle after its original assembly. An aftermarket V2V communication safety device with safety applications is capable of providing advisories and warnings to the driver of a vehicle similar to those the alerts provided by an OEM-installed V2V device. These devices, however, may not be as fully integrated into the vehicle as an OEM device, and the level of connection to the vehicle can vary based on the type of aftermarket device itself. For example, a "self-contained" V2V aftermarket safety device could only connect to a power source, and otherwise would operate independently from the systems in the vehicle. The self-contained device would include a DVI for advisories and warnings. Aftermarket V2V devices can be portable devices inserted into a fixed cradle in the vehicle (e.g., cell phones with applications), or added to a vehicle at a vehicle dealership or by authorized dealers or installers of automotive equipment.

Vehicle Awareness Devices (VADs): The VAD is the simplest design. It only transmits a BSM to nearby vehicles. A VAD does not have any safety applications or a driver interface; and it cannot provide any advisories or warnings to a driver. Installing these devices on existing vehicles could be an attractive option for fleet operators, rental agencies, or vehicle owners who could see benefit in signaling the presence of their vehicles to V2V-equipped vehicles and thus potentially avoiding crashes. Installation of VADs could increase deployment of V2V systems across the fleet as a whole, and thus potentially could increase the benefits for early adopters of this technology.

During the Safety Pilot Model Deployment project over 3,000 vehicles have been equipped with a range of these devices—sixty-four vehicles were equipped with integrated OEM solutions (CAMP-developed device) that have been fully integrated into the vehicles, 300 vehicles had aftermarket technology installed, 19 heavy vehicles (16 trucks and 3 transit buses) were retrofitted with equipment, and 2,850 vehicles were outfitted with vehicle-awareness devices, which can transmit the BSM to other vehicles, but cannot receive information needed to alert the driver. Many of these systems had internal components designed and built by a number of different manufacturers and suppliers, illustrating interoperability across makes and models. With these different devices operating together, as a system, providing alerts and advisories to drivers, USDOT has generated a representation of how a fully functional V2V system might work. Additionally, the results on interoperability provide USDOT, OEMs, and State and local agencies with a basis for planning for "day-one" benefits knowing that a range of devices can be deployed together to start the process of transforming safety, mobility, and environmental performance.

E.I.B. External V2V and V2I Components

To enable communication between vehicles and roadside equipment and support V2I applications, DSRC must be integrated with existing traffic equipment (e.g., traffic signal controllers or backhaul connections to Traffic Management Centers (TMCs)). All of these components are considered "roadside equipment" (RSE) or "roadside infrastructure". These terms describe categories of infrastructure components that include DSRC-based roadside units (RSUs), which are installed as part of the RSE. DSRC RSUs are essentially devices/radios and act as data processors that facilitate communication between vehicles, other devices, and transportation infrastructure by exchanging data over the 5.9 GHz DSRC band.

As illustrated in Figure E-2 on the next page, examples of roadside infrastructure components include:[207]

- **Infrastructure communications equipment**: Including a DSRC radios (RSU) that allow the V2I safety infrastructure applications platform and other Connected Vehicle services (e.g., security services) to communicate wirelessly at the necessary latency and network attach time to exchange information with other DSRC-equipped devices. Roadside communications equipment also includes other elements not requiring low latency wireless transmission that link the infrastructure applications platform with local supporting equipment such as traffic signal controller units, and linking to off-site, back office elements such as Traffic Management Centers and centers supporting Connected Vehicle security. This is known as backhaul communications.

---

[207] Based on: http://ntl.bts.gov/lib/48000/48500/48527/ED89E720.pdf, page 79. (Last accessed June 2015)

- **Applications:** An infrastructure application platform that hosts V2I safety applications, receives information relevant to those applications from roadside detectors and traffic signal controller units, and communicates with the infrastructure communications equipment.

- **Data Sensors and Other Equipment:** Infrastructure data equipment, such as traffic signal controller units, weather detectors, pedestrian defectors, and traffic detectors that provide information to the V2I safety application platform.

- **Driver-Vehicle Interface (DVI):** A DVI that—based on the data analysis conducted by the V2V and V2I software applications—provides a warning to the driver through infrastructure-based features such as a changeable, dynamic message sign that is compliant with the Manual on Uniform Traffic Control Devices to alert drivers to upcoming hazards or dangerous conditons.

Figure E-2 illustrates one example of the interaction between vehicle- and infrastructure-based components. There will be other ways to perform these functions, or additional functions that may be added, but this illustrates a standard configuration at many V2I deployments.

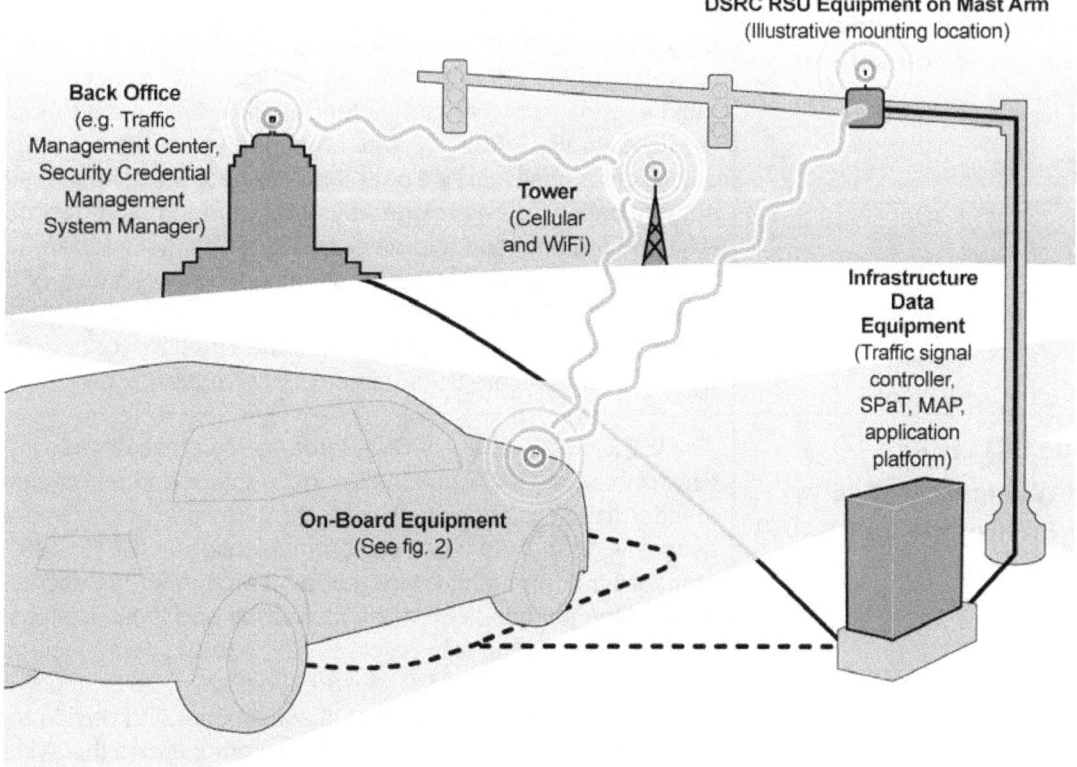

Figure E-2.   Basic External V2V and V2I Roadside Infrastructure Components for a Typical Deployment (Image Source: USDOT)

E.II. Standards

An important part of ensuring interoperability is making sure that DSRC functions properly, exchanges information the same way every time, and uses standardized messages. Cooperative system standards were developed specifically to support V2V and V2I wireless interfaces. They establish a wireless link for V2V and V2I communications (IEEE 802.11p), establish protocols for information exchange across the wireless link (IEEE 1609.x), and define message content for communicating specific information to and from equipment and devices via DSRC (SAE J2735 and SAE 2945.x) or other communications media. Each of the cooperative system standards facilitates some part of DSRC operation. Table E-1 describes the fundamental Cooperative System Standards for V2V and V2I communications.

Table E-1. Cooperative System Standards for V2V and V2I Communications

Cooperative System Standards	Description
IEEE 802.11p-2012 **Medium Access Control and Physical Layer Specifications for WAVE**	IEEE 802.11 is a set of standards that specify the physical layer for implementing wireless local area network (WLAN) using Wi-Fi bands. Only certain parts of the standard are required for implementing DSRC operating at 5.9 GHz for V2V communications. To accommodate the rapid exchange of trajectory information between vehicles traveling at high speed, IEEE 802.11p was amended to enable operation without setting up a basic service set. It allows security services, such as authentication, to be provided by other standards. It describes adjacent channel and alternate adjacent channel interference criteria and transmission masks corresponding to requirements of the FCC rules for DSRC. The entire standard applies to V2V and V2I communications, because it defines the structure for how devices should communicate using the 5.9 GHz frequency band,.
IEEE P1609.0/D5.8 **Guide for Wireless Access in Vehicular Environments (WAVE) Architecture**	The IEEE 1609 Family of Standards for Wireless Access in Vehicular Environments (WAVE) define an architecture and a complementary, standardized set of services and interfaces that collectively enable secure V2V and V2I wireless communications. IEEE 1609.0 is not a standard, but an architecture guide. It provides the descriptions of each of the full-use IEEE 1609 standards and their relationships to other relevant standards (such as IEEE 802.11), and includes guidance on how they should work together. The protocol architecture, interfaces, spectrum allocations, and device roles are all described. The guide is intended for organizations that will implement DSRC, such as state departments of transportation, automobile and original equipment manufacturers, aftermarket equipment makers, application developers, and standards developers.

Cooperative System Standards	Description
IEEE 1609.2-2013 **Security Services for Applications and Management Messages**	This standard describes security services for WAVE management messages and application messages designed to meet these goals. This standard was intended to be used primarily for DSRC. The safety-related content of WAVE applications, and particularly vehicle safety applications, makes it necessary to protect messages from attacks such as eavesdropping, spoofing, alteration, and replay. Recipients of safety messages have to be assured that the messages they receive are authentic and are sent by a source authorized to transmit those messages. Additionally, the fact that the WAVE technology may be implemented in communication devices in personal vehicles as well as in other portable devices whose owners may have some expectation of privacy means that the security services need to be designed to avoid, for example, revealing personal, identifying, or linkable information to unauthorized parties in systems where PII may be involved.
IEEE 1609.3-2010 **Networking Services**	IEEE 1609.3 specifies how various message types (e.g. WAVE Short Messages, WAVE Service Advertisements, and WAVE Routing Advertisements) are assembled, packaged, and handled between an application and IEEE 1609.4 for transmission or upon reception. It describes how to build, route, process, and interpret WAVE low latency messages, as well as messages based on other well-known protocols such as the User Datagram Protocol (UDP) and Internet Protocol Version 6 (IPv6). The standard includes information on which messages go on the control channel, which messages go out on the service channels, advertising specific services, authenticating the messages, accessing applications hosted on an external network (e.g., the Internet), and methods for how this can be accomplished.
IEEE 1609.4-2010 **Multi-Channel Operations**	This standard describes multi-channel radio operations for WAVE. It is used in conjunction with other IEEE 1609 standards and IEEE 802.11-2012 to implement DSRC communications in the 5.9 GHz frequency band. WAVE operates using IEEE 802.11 outside the context of a basic service set. In order to implement functions such as user priority access to the media, routing data packets on the correct channel with the desired transmission parameters, and the ability to coordinate switching between the control channel and service channels, additional functions are required between the IEEE 802.11 medium access control (MAC) and the Logical Link Control (LLC). This standard specifies how these functions are implemented.
IEEE 1609.12-2012 **Identifier Allocations**	IEEE 1609.12 describes the format and use of the provider service identifier (PSID), and indicates identifier values that have been allocated for use by WAVE systems.

Cooperative System Standards	Description
SAE J2735, Version 2 **DSRC Message Set Dictionary**	The SAE J2735 standard specifies message sets, data frames, and data elements that make up messages/dialogs specifically for use by applications intended to utilize the 5.9 GHz DSRC for WAVE communications systems. The messages for V2V safety applications are defined in SAE J2735 as the Basic Safety Message (BSM) parts 1 and 2. Other parts of SAE J2735 define the message sets for other ITS applications, such as weather and mobility.
SAE J2945.1, Version 1 **DSRC Minimum Performance Requirements**	The SAE J2945.1 standard specifies the minimum communication performance requirements of the DSRC message sets and the necessary BSM data elements to support V2V safety applications.

The ITS Joint Program Office's Standards Program funds and manages ITS cooperative system standards efforts and international harmonization (harmonization efforts are discussed in Appendix H of this report). The content of these standards is developed collaboratively with contributions from diverse stakeholders. The CAMP research has resulted in significant contributions to many of the standards described above.[208]

The current status of these standards are described in the DSRC Technical Milestones, Appendix D of this report.

---

[208] Specifically, VSC-A and CAMP have contributed to the development of SAE J2735 (DSRC Message Set Dictionary); SAE J2945.1 (DSRC BSM Minimum Performance Requirements); IEEE 1609.0 (Architecture); IEEE 1609.2 (Security Services); IEEE 1609.3 (Networking Services); IEEE 1609.4 (Multi-Channel Operation); IEEE 1609.12 (Identifier Allocations); and IEEE 802.11p (Wireless Access in Vehicular Environments (WAVE)).

Appendix F   Definition of Connected Vehicle Safety Applications–V2V and V2I

F.I   V2V Safety Applications

The V2V safety applications address common rear-end, opposite direction, junction crossing, and lane change crash scenarios, as described below. [209]

Table F-1. Description of the V2V Safety Applications

Crash Type	Safety Application	Description
Rear-End	Forward Collision Warning (FCW)	Forward Collision Warning (FCW): Warns the driver of one vehicle in case of an impending rear-end collision with another vehicle ahead in traffic in the same lane and direction of travel.
Rear-End	Electronic Emergency Brake Light (EEBL)	Emergency Electronic Brake Light (EEBL): Warns the driver of a vehicle in the case of another vehicle that is braking hard up ahead in the flow of traffic. The braking vehicle does not necessarily have to be in the direct line of sight of the following vehicle, and can be separated by other vehicles
Opposite direction	Do Not Pass Warning (DNPW)	Do Not Pass Warning (DNPW): Warns the driver of one vehicle during a passing maneuver attempt when a slower moving vehicle, ahead and in the same lane, cannot be safely passed using a passing zone that is occupied, or will soon be occupied, by vehicles in the opposite direction of travel. The application may also provide the vehicle driver an advisory warning that the passing zone is occupied when a passing maneuver is not being attempted.
Opposite direction	Left Turn Assist (LTA)	Left Turn Across Path/Opposite Direction (LTAP/OD): Warns the driver of one vehicle when they are entering an intersection where another vehicle (traveling in the opposite direction) is conducting a left turn maneuver across the path of the original vehicle.

---

[209] Reference: "Vehicle-to-Vehicle Communications: readiness of V2V Technology for Application, DOT HS 812 014, pg. 136, at: http://www.nhtsa.gov/staticfiles/rulemaking/pdf/V2V/Readiness-of-V2V-Technology-for-Application-812014.pdf. (Last accessed June 2015)

Crash Type	Safety Application	Description
Junction crossing	Intersection Movement Assist (IMA)	Intersection Movement Assist (IMA): Warns the driver of one vehicle when it is not safe to enter an intersection due to high collision probability with other vehicles at controlled and uncontrolled intersections.
Lane change	Blind Spot Warning (BSW) + Lane Change Warning (LCW)	Blind Spot Warning (BSW) + Lane Change Warning (LCW): Warns the driver during a lane change attempt if the blind spot zone into which the driver intends to switch is, or will soon be, occupied by another vehicle traveling in the same direction. The application also provides the driver with advisory information that another vehicle in an adjacent lane is positioned in the original vehicle's "blind spot" zone when a lane change is not being attempted.
Lane change, Front-to-Side	Vehicle Turning Right in Front of Bus Warning (VTRW)	Vehicle Turning Right in Front of Bus Warning (VTRW): Warns a bus driver of the presence of a vehicle attempting to go around the bus to make a right turn as the bus departs from a bus stop.

F.II. V2I Safety Applications

V2I safety applications under development include applications for commercial freight operators and transit agencies. V2I applications complement the V2V safety applications by addressing crash scenarios that the V2V program cannot address or that could be addressed more efficiently with low levels of penetration of DSRC-equipped light vehicles. Table F-2 provides a list of V2I potential safety applications:

Table F-2. V2I Safety Applications

Title	V2I Safety Applications
Red Light Violation Warning (RLVW)	Based on vehicle speeds and distances to intersections, this technology will provide in-vehicle alerts to drivers about potential violations of upcoming red light.
Curve Speed Warning (CSW)	If a driver's current speed is unsafe for traveling through an upcoming road curve, this technology will alert the motorist to slow down.
Stop Sign Gap Assist (SSGA)	This technology will assist drivers at stop-sign-controlled intersections via vehicle gap detections, alerting motorists when it is unsafe to enter intersections.
Stop Sign Violation Warning (SSVW)	Based on vehicle speeds and distances to intersections, this technology will provide in-vehicle alerts to drivers about potential violations of upcoming stop signs.
Reduced Speed Zone Warning (RSZW)	This technology will assist drivers in work zones, by issuing alerts to drivers to reduce speed, change lanes, and/or prepare to stop.

Title	V2I Safety Applications
Spot Weather Information Warning (SWIW)	This technology will provide in-vehicle alerts or warning to drivers about real-time weather events and locations, based upon information from DSRC RSU connections with Transportation Management Center (TMC) and other weather data collection sites/services.
Railroad Crossing Violation Warning (RCVW)	This technology will assist drivers at controlled railroad crossings via DSRC RSU connections with existing train detection equipment, alerting motorists when it is unsafe to cross the railroad tracks.
Oversize Vehicle Warning (OVW)	Drivers of oversized vehicles will receive an in-vehicle alert to take an alternate route or a warning to stop, based upon information from DSRC RSU connections to infrastructure at bridges/tunnels.
Pedestrian in Signalized Crosswalk Warning (PCW)	Warns a bus driver if a pedestrian is in the crosswalk of a signalized intersection and in the intended path of the bus when the bus is making a right or left turn.

As noted previously, V2I safety applications are dependent on the cooperation of infrastructure and vehicle components to achieve the system's operational objectives. The applications listed in Table F-2 require broadcast messages to support their functionality that include, but are not limited to, the following:

- Signal Phase and Timing (SPaT): contains information and current status on the phase and timing of all the signals for each approach in the intersection. This message, together with the intersection geometry information (MapData or "MAP" described below), will enable the vehicle to determine which signal indication applies to it and use this information for determining whether a warning is warranted.

- MapData (MAP)
    - Road geometry, including curve radius and super elevation.
    - Intersection Geometry Information: Intersection information, including intersection ID, road/lane geometry for all approach roads (e.g., geometric intersection design or "GID"), location of stop lines, and lane numbering scheme associated with movements

- Positioning correction (optional): global positioning system (GPS) positioning correction information for the intersection that the vehicle may use to improve its estimate of location within the intersection.

- Road surface information and other weather-related data if available (Optional): information about the road surface coefficient of friction at the intersection, weather related data such as dew point, temperature, visibility, and rain, which may assist the vehicle components in adjusting warning timing to account for variations in stopping distance.

Appendix G Proposed Certification Path

USDOT has developed a four-layered approach to device certification (described below).[210] USDOT is proposing that a basic device must be certified at layers one (1) and two (2). An application that resides on a basic device must be certified for its functional and communications capabilities requiring all four layers. This structure allows for multiple applications to be hosted on one device and use lower layer certifications (layers 1, 2, and 3). Each layer is further described below.

1. **Layer 1: Environmental Abilities:** This layer certifies that a device is suitable for the environment in which it will be used (temperature, vibration, weather, etc.). Environmental suitability certification is necessary because the device's contribution to the application or system is lost if the basic device fails prematurely. As listed below, various industry segments will likely have specific requirements in this area. Self-certification is potentially appropriate at this level, as it is believed that responsibility to customers will enforce conformance.

   i.   Automotive component Tier 1 suppliers are familiar with environmental certification based on specifications from their Original Equipment Manufacturers (OEM) customers.

   ii. Device specifications will inform roadside device manufacturers on appropriate requirements for their products.

   iii. Aftermarket device makers will need to meet environmental requirements suitable for the cost/grade of their products.

Conformance with Layer 1 is considered necessary, but not sufficient, to obtain security credentials. A Security Credential Management System (SCMS) would need to maintain a list of conforming devices by make and model as defined by industry standards.

2. **Layer 2: Communication Protocol Abilities:** This layer certifies that the basic device conforms to the communications protocol standards that govern transfer of message payloads (e.g. radio service interoperability for DSRC, message payload format consistency). The basic device must meet a specific configuration of the specified standards to ensure interoperability.

---

[210] Summarized from the August 2013 Request for Information at: https://www.fbo.gov/index?s=opportunity&mode=form&id=7dc0a89ceb599a34ae7e4882aaad2517 (Last accessed June 2015)

USDOT and system contributors will need to work collectively to arrive at an understanding of the specific requirements for this level. The ability to perform certifications—either self-certifications or third-party lab certifications—will need to be established by the system contributors. Federal Communications Commission (FCC) license requirements and application message definition requirements will make conformance necessary:

    i.   Radio service operation will likely be done by specialized labs.

    ii.   Message payload construction could be done by specialized labs or specific users with the necessary expertise.

Conformance at Layer 2 is considered necessary, but not sufficient, for device models (combination of hardware and operating system and communication protocol stack software) to obtain security credentials. An SCMS would need to maintain a list of conforming devices by make and model as defined by government regulations (FCC, NHTSA, and FMCSA) or industry standards.

3. **Layer 3: Interface Abilities:** This layer is applicable to one or more applications and certifies that the data elements, of a message payload include the appropriate payload content and that it is formatted properly. In short, the certification verifies that both the syntax and contents being shared across the interface is useful to an application. This layer defines operating conditions and criteria for meeting accuracy and performance requirements. Product development processes and industry or governmental regulations could enforce conformance.

    i.   OEMs may conduct this level of certification with initial samples from device makers.

    ii.   Third party testing and certification organizations could also certify at this layer.

Conformance at Layer 3 is considered needed for a specific service or application software component to get access to security credentials. A SCMS would need to maintain a list of approved service and application components by author and revision level as defined by government regulations or industry standards.

4. **Layer 4: Overall Application Abilities:** This layer is the most difficult to certify. In order to ensure that all specifications are verified, an entity must bring multiple components together to see if the whole system works. One example of this certification layer can be seen in Crash Avoidance Metrics Partnership (CAMP) application developers' role to integrate vehicles and infrastructure communications in the Safety Pilot Model Deployment. Similar overall interest groups will need to do this for all applications.

    i.   Industry associations may do this layer of certification prior to introducing a new application and at regular intervals thereafter based on modifications to an application.

    ii.   Not all applications will require certification at this layer, but which ones may be excluded has not yet been determined.

    iii.   The mandated DSRC technologies could be accompanied by specific objective test procedures with specific pass/fail criteria in the form of FMVSS certification testing.

The results of Layer 4 testing would confirm the results used in Layers 1, 2, and 3 and verify the system level functions. An SCMS would need to maintain a list of the standards by name and revision level used at each certifying event. Those standards would have to be referenced by device makers (Layers 1 and 2) and application software developers (Layers 3 and 4) who want their products listed as suitable for receiving security credentials.

Appendix H International Uses of DSRC

H.I. Differences between the US regional vision and other regions[211]

H.I.A. Comparison of US to European Union (EU)

The US approach focuses on a core set of crash-critical V2V safety applications. In previous research conducted by USDOT under the Vehicle Infrastructure Integration (VII) Program, the major focus was V2I applications and establishing an infrastructure. The shift in primary focus to vehicle-based V2V applications facilitates implementation of ITS safety technologies without the costly infrastructure implemented through state and local government investment while achieving safety benefits at overall lower costs. While the EU has defined crash-critical safety applications as well, the priority in the EU is driver safety advisories (not safety-critical warnings), driver support messages (such as eco-driving), and commercial applications such as insurance.[212] The breadth and content of EU applications, including mobility applications, reflects their market driven approach, whereas the V2V safety focus in the US reflects the potential for reducing crashes.[213] In the EU standards development activities encompass a broader set of applications while USDOT is primarily focused on developing standards to support V2V crash avoidance applications.[214] Release 2 of the European Telecommunications Standards Institute (ETSI) standards, planned for 2017, will focus on crash avoidance.

European carmakers have committed to begin introducing DSRC systems in 2015 and it is likely that initial European introductions would be on high-end vehicles and/or newly re-designed vehicle models; a different approach than requiring DSRC on all vehicles. While initial introduction in Europe could come much sooner than the US, the number of equipped vehicles could grow faster after the initial start in the US, if the US pursues a DSRC mandate for all new vehicles.

The current European model includes infrastructure costs that are not envisioned in the initial stages of V2V implementation in the US

---

[211] Summarized from NHTSA: "Vehicle-to-Vehicle Communications: Readiness of V2V Technology for Application, DOT HS 812 014, at: http://www.nhtsa.gov/staticfiles/rulemaking/pdf/V2V/Readiness-of-V2V-Technology-for-Application-812014.pdf (Last accessed June 2015)

[212] Global V2X Deployment: Contrasts with U.S. Approach, at 35 (Bishop, Jan. 21, 2013)

[213] Ibid.

[214] Ibid.

In terms of spectrum allocation, the US allocation calls for seven channels of 10 MHz each (a total of 75 MHz of spectrum located in the 5.85 to 5.925 GHz frequency band), with one channel designated as a control channel and one channel exclusively for safety.[215] The EU allocation calls for the 5.875-5.905 MHz band to be designated for safety related ITS functions with three 10 MHz channels, including the possibility of two additional channels being granted in the future. No control channel exists in the EU approach.

H.I.B   Comparison of US to Asia

In Asia, Japan and Korea are most active in DSRC development, with Japan leading. In both countries, the initial focus is on adapting the Electronic Toll Collection system operating at 5.8 GHz. The Japanese government has deployed 5.8 GHz "ITS Spots," which communicate with electronic toll tags to offer limited V2X safety capabilities, as well as mobility and convenience services. Additionally, some Japanese automotive OEMs (mainly Toyota) are actively supporting the deployment of V2X using 760 MHz communications.

Japan appears likely to proceed with a two-band solution, and suppliers have prototyped transceivers covering both bands. Deployment of 760 MHz systems could come as soon as later 2015/early 2016.[216]

Similar to the approach in Europe, deployment in Japan is mostly market-driven, with the government leading to provide initial roadside capability in the case of the 5.8 GHz system, and some OEMs pushing for the 760 MHz system for V2V crash avoidance.

The Japanese 5.8 GHz system is not compatible with the IEEE 802.11p protocol used in the US and Europe, due to a Japanese law requiring legacy protocols. At the security level, there are advocates of using IEEE 1609.2 as the security framework, which would be compatible with the US and Europe, but this has not yet been decided. Harmonization of probe data message sets is currently underway between Japan and the US. Development of message sets in Japan is not yet complete but appears to be toward the harmonized version of the (US) BSM and the (EU) Cooperative Awareness Message (CAM) / Decentralized Environmental Notification Message (DENM) message sets.

In China, this band is reserved for potential ITS use as well although no information is available indicating any interest from China for ITS applications in the 5.9 GHz DSRC band.

---

[215] For details, see: http://www.drive-c2x.eu/tl_files/publications/3rd%20Test%20Site%20Event%20TSS/1%20DRIVE%20C2X%203rd%20Test%20site%20event_Lan%20Lin_Technology_20130613.pdf. (last accessed July 2015)

[216] See information on OEM planned production releases with this technology at: http://www.traffictechnologytoday.com/news.php?NewsID=64353 (last accessed July 2015)

There have been indications that Korea seeks to shift to 5.9 GHz to be more compatible internationally, but no announcements have been made.

H.I.C Australia

Australia is moving forward with their planning for Connected Vehicle environments but they face a challenging issue—with no remaining native vehicle manufacturing, all vehicles will be imports going forward. Thus, if the Japanese, US, and European decisions on technologies, spectrum use, and standards differ enough to impact the foundational systems, Australia may find that is will have to choose one approach versus any other. This will constrain market choices for Australians as well as constrain opportunities for OEMs and suppliers. This situation highlights the significance of harmonization efforts, which Australia has joined informally and is seeking to formalize its participation going forward.

H.II. Harmonization of International Standards and Architecture around the Vehicle Platform

Because the automobile industry is global, it is critical to reduce barriers to standardization and achieve a broad agreement on harmonization that can benefit both the public and the motor vehicle industries. The objective of the Standards Harmonization research program is to work with the international standards community to harmonize standards and architecture to increase vehicle connectivity. Harmonization facilitates interoperability between products and systems, which can benefit transportation management agencies, vehicle manufacturers, equipment vendors, and others. By overcoming institutional and financial barriers to technology harmonization, stakeholders could realize lower life-cycle costs for the acquisition and maintenance of systems. Efforts under this research program include collaboration with standards development organizations, original equipment manufacturers, and other stakeholders to seek agreement and provide appropriate incentives.

Joint standardization of Connected Vehicle systems (V2V and V2I) is a core objective of the EU-US cooperation on ITS. In 2012, two Harmonization Task Groups (HTGs) worked in parallel on analyses of security standards (HTG1) and communications standards (HTG3). The HTG1 analysis indicated that the initial set of security standards, developed by IEEE, was reasonably well harmonized. However, the group found that gaps still exist and must be addressed before large-scale deployment occurs. Some of these gaps are related to policy, and an HTG6 effort was launched in 2014 to address gaps. The HTG3 analysis involves the current differences between the EU and US approach to standardizing communications at 5.9 GHz for V2V and V2I safety-of-life and property communications. The HTG3 identified differences that presented interoperability challenges and offered suggestions for harmonizing these differences to achieve a single global standard.

An outcome of the HTG1 and HTG3 work was recognition of the need to harmonize security policies and standards. To meet this need, a third HTG (HTG6) was established to explore and find consensus on management policies and security approaches for cooperative ITS. This could be assessed across international, regional, and local levels to determine optimal candidate guidelines for policy areas. HTG6's intent is to identify where harmonization is desirable by exploring the advantages and limitations of global versus local security policy alternatives, including economic benefits. The task group is identifying the largest set of common approaches and the benefits for commonality and identifying those policies and approaches that need to differ regionally and the reasons for divergence.

The final deliverable of HTG6 will be an end-to-end security policy framework and will be available in Summer 2015 in draft form for stakeholder feedback. Final, revised documents will be provided by the end of 2015.[217]

HTG1 and HTG3 group members comprise a small group of international experts who worked together intensively with co-leadership provided by the EC DG-CONNECT and USDOT. These experts were chosen from among the editors of many of the current cooperative ITS standards in the different SDOs providing direct linkages into those SDO activities, and representatives of the EU and USDOT and the VIIC plus an observer from Japan. HTG6 uses a similar model of gathering a small group of experts; is co-led by the EC DG-CONNECT, USDOT, and Transport Certification Australia (TCA) plus observers from Canada and Japan.

HTG1 and 3 results provide guidance to the SDOs for actions to be taken that raise the assurance of security interoperability of deployed equipment. Vehicle connectivity through harmonization of standards and architecture will reduce costs to industry and consumers in that hardware and/or software development costs will be spread over a larger user base, resulting in reduced unit costs. Differences between vehicles manufactured for different markets will also be minimized, allowing private-sector markets to have a greater set of global opportunities.

---

[217] HTG6 documents will be posted at this link: http://ec.europa.eu/digital-agenda/en/news/harmonized-security-policies-cooperative-intelligent-transport-systems-create-international (last accessed November 2015)

Appendix I   Spectrum Operations

I.I.   Using the DSRC Spectrum

This section describes the spectrum allocation and operations needed to enable access by the technologies that support the applications described in Section 2.I.[218] The FCC has allocated 75 MHz of wireless spectrum for the DSRC. This spectrum is divided into seven non-overlapping 10 MHz channels, plus a 5 MHz guard band at the beginning of the frequency range. The FCC band plan for this spectrum specifies particular usage, power limits, etc. for these channels as shown in Figure I-1 on the following page.

DSRC radio units operate on one frequency (or "channel") at a time. This is similar to the way the AM/FM radios in vehicles today operate. These radios can receive one station or another depending on how they are tuned (tuning being the act of shifting signal reception from one radio frequency to another), but they do not receive clearly when they are between stations, and cannot be tuned to more than one frequency at a time.

The current V2V operation utilizes two radios. One radio operates in continuous access mode, always tuned to channel 172 for dedicated crash avoidance safety-of-life applications, including broadcast of the BSM. The other radio operates in channel-switching mode, using channel 178 as a control channel to manage channel switching[219] to support messages on other channels related to security-services and other services/applications, such as mobility or environment.

---

[218] Summarized from NHTSA: "Vehicle-to-Vehicle Communications: Readiness of V2V Technology for Application, DOT HS 812 014, pages 109-114, at: http://www.nhtsa.gov/staticfiles/rulemaking/pdf/V2V/Readiness-of-V2V-Technology-for-Application-812014.pdf (Last accessed June 2015)

[219] Channel switching is the utilization of a dedicated channel to route incoming messages to multiple "service" channels that utilize the incoming information. This method allow for a single radio to be used to support multiple functions.

Researchers initially attempted to use channel 178 as both a "control" channel[220] and for transmission of the BSM, but this unduly restricted BSM transmission, potentially hindering safety. It was thought that a channel-switching mode could be used on a single radio to support the BSM as well as use the other channels for other messages, because the channel switching mode would cause the BSM transmissions to switch from channel 178 to some other channel. However, because a radio can only transmit or receive on a single channel at a time, channel switching only solves part of the problem—the radio still has to alternate between the BSM and the other necessary messages, which degrades its ability to support the BSM, as described below.

The sections that follow explain the alternative modes of DSRC operation and how the research indicated the need to implement a dedicated channel for the BSM.

I.I.A Channel Switching Mode

The DSRC will have to switch from one channel to another in order to transmit and receive messages on different channels, which may be needed in order to perform different functions necessary for V2V communications.

Transmission latency is a critical aspect of V2V communications, because BSM transmissions need to be received in a timely manner to warn drivers of potential dangers in time for them to react. If DSRC is switching from one channel to another, it may experience a time lag as the next channel is being "picked up," which may potentially affect receipt of important transmissions. The IEEE 1609.4 standard[221] divides time for purposes of DSRC transmission into 100 millisecond (ms) sync intervals (the equivalent of 10Hz). The sync intervals are then sub-divided into a Control Channel (CCH) interval and a Service Channel (SCH) interval. A time division mechanism is defined for a device to switch between the CCH and a SCH every 50 ms to transmit and/or receive DSRC messages.

As shown in Figure I-2 below, Channel 178 is designated as the CCH. It was originally envisioned that all vehicle and roadside units accessing this spectrum would use the control channel to determine what information was available on other channels, and then switch to the other channels as needed to access the information.[222]

---

[220] The control channel "tells" the radio which channel to "listen" to for specific information as well as transmitting that same information when the device is ready to transmit information.
[221] For more information, *see:* VSC-A Final Report: Appendix Volume 2.
[222] Ibid.

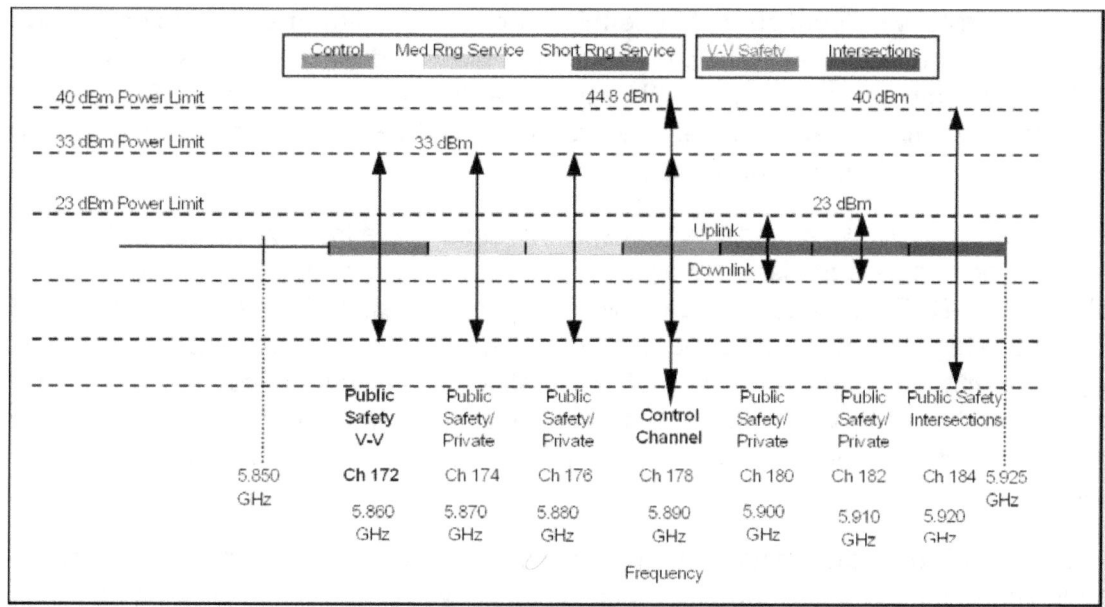

Figure I-1. DSRC Channel Assignment Image: USDOT/CAMP

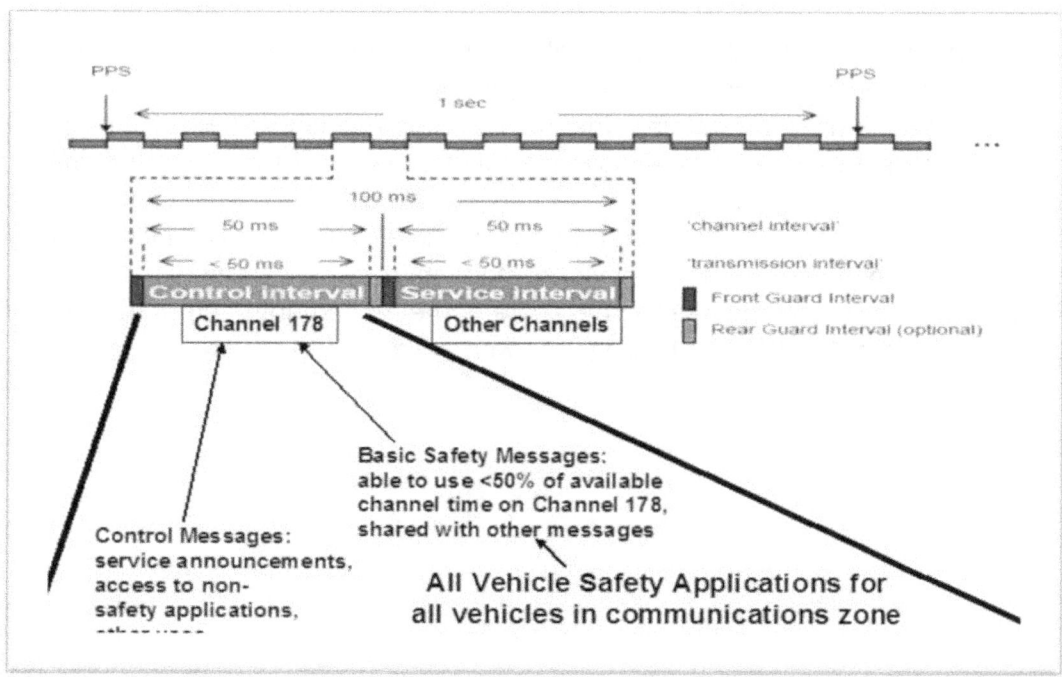

Figure I-2. Time-Division Channel Usage Image: USDOT/CAMP

For testing, vehicles were configured to use this channel switching operation to send and receive BSMs on the CCH during the CCH interval. This would allow vehicles to participate in non-V2V safety communications on a SCH during the SCH interval for other DSRC services. While this safety communication model is not required by IEEE 1609.4 or any other standard, it was considered as the baseline approach.

The above approach allows a single-radio vehicle to participate in V2V safety by exchanging BSMs with its neighbors and also to avail itself of DSRC services that are offered during SCH intervals (e.g., by DSRC RSUs). This capability is especially attractive as part of an initial DSRC deployment strategy to boost market penetration. One of the main disadvantages is that safety messages are effectively limited to the CCH interval, and thus channel congestion may be a concern and lead to a probability that two or more packets could "collide" due to overlapping transmissions at high channel loads. As explained below, there are options for mitigating these limitations.

NHTSA and CAMP are performing research on two different congestion mitigation algorithms and continuing to test channel load thresholds, especially under conditions of higher vehicle densities. These algorithms were specified under the system requirements and units were fielded with these predetermined algorithms. They worked well and predictably under all test scenarios.

As a point of reference, Figure I-3 shows the interchange between I-495 and I-66 outside of Washington, DC. This interchange contains two express lanes and four regular lanes for I-495 running north and south and passing underneath I-66, which has three lanes running east and west. When off ramps are added, this leads to a total of 22 lanes of traffic in a 300 meter radius. In grid-lock conditions, where assuming an average car takes 24 feet of lane space, this interchange can have over 800 vehicles in range of a single radio. It must be noted, however, that during gridlock when vehicles are at a virtual standstill, the probability of crashes is significantly reduced although the need to be aware of the movement of other travelers (for instance, motorcycles or bicyclists or pedestrians when allowed) remains.

Figure I-*3*. I-495 and I-66 Interchange Image: Courtesy of CAMP

I.I.B. Multi-Channel Operation versus a Dedicated Safety Channel

Having two radios, one of which is always tuned to the dedicated safety channel, may help to avoid the need for channel switching and enable the vehicle to broadcast and receive BSMs the entire time it is in operation. Tests performed by CAMP in 2010 formed the basis for a comparison of the two approaches. The results were provided to the FCC in April 2011 to offer greater detail on how Channel 172 ought to perform under operational conditions (see section 3.I, Policy Milestones for more information about this meeting). As a result of these tests, and with FCC's concurrence, industry has had a stable definition from which to develop devices and applications with the appropriate protocols.

I.I.C. Interoperability performance requirements

In a Connected Vehicle environment DSRC, GPS, and other system components will likely require certification confirming that all devices meet performance requirements, in order to ensure interoperability. Appendix D.II: DSRC: Technical History provides the history of how these performance requirements were developed, and contains a table that illustrates the level of maturity and "readiness" of each requirement.

The following factors were taken into account in developing performance requirements:

- Safety application characteristics;
- The transmitting power a DSRC radio could provide; and its receiving ability at a given area with a given transmitting power;
- The language vehicles speak when they communicate with one another;
- The language used for communication between vehicles and DSRC RSUs;

- Information necessary to be included in the BSM;

- Information necessary to be included in the communication between vehicles and infrastructure;

- Media devices used to carry messages when they communicate with one another, and media devices used to carry messages when they communicate with DSRC RSUs;

- Basic Safety Message data accuracy needs;

- Error tolerance and error correction capability (considering potential distortion) of over-the-air signals being received by OBE;

- Capability of the system to accommodate all communication within a given area of coverage and for a given number of vehicles (DSRC channel congestion mitigation);

- Method of synchronizing communication system network;

- The method of verifying and validating messages from other vehicles; and verifying and validating messages from other ECUs in a vehicle itself;

- Security scheme to protect data communications, initiate and ensure trusted key establishment, support credential management, and protect Personally Identifiable Information (PII);

- Physical security to protect security components and elements that will be essential pieces of establishing and sustaining the network trust at the Infrastructure side; and

- Physical security to protect security components and elements that will be essential pieces of establishing and sustaining the network trust on the on-board DSRC devices.

In a set of iterative research periods, these requirements have been defined and refined, and broken up into both functional (high-level) requirements and performance (detailed) requirements.[223] The V2V-Interoperability Report contains requirements for the on-board equipment (DSRC radio, GPS receivers, and processors). Some of the requirements that were developed during these projects have been worked into a number of IEEE and SAE standards and/or device specification documents. For instance, the performance requirements that were used and implemented in the specification documents for the VADs and ASDs during the Safety Pilot Model deployment were developed directly from this research.

---

[223] The critical requirements can be found in sections 4 and 5 of System Requirements Description, 5.9 GHz DSRC Vehicle Awareness Device Specification, Version 3.6 (Jan. 25, 2012) *at:* http://www.its.dot.gov/newsletter/VAD%20Specs.pdf (Last accessed June 2015).

Appendix J   National Research Council Comments

In accordance with Congressional intent, the DSRC Report to Congress report was reviewed to incorporate comments received from the National Research Council (NRC) and an independent third party and source of subject matter expertise that was managed by the NRC to validate conclusions. The NRC submitted a Letter Report with a total of 21 comments and recommended revisions. USDOT's ITS Joint Program Office, working collaboratively with the Volpe National Transportation Systems Center, has responded to each comment by revising the *2015 Report to Congress* as needed.

The NRC's Letter Report is attached in its entirety, beginning on the following page. It can also be accessed at: http://onlinepubs.trb.org/onlinepubs/reports/DSRC_April_28_2015.pdf.

Letter Report

Review of the *Status of the Dedicated Short-Range Communications Technology and Applications [Draft] Report to Congress*

Transportation Research Board
Of The National Academies
2015

TRANSPORTATION RESEARCH BOARD
OF THE NATIONAL ACADEMIES

April 28, 2015

The Honorable Anthony R. Foxx
Secretary
U.S. Department of Transportation
1200 New Jersey Avenue, SE
Washington, D.C. 20590-9898

Dear Secretary Foxx:

As required in Section 53006 of P.L. 112-141 (Moving Ahead for Progress in the 21st Century, dated July 6, 2012), Congress directed the U.S. Department of Transportation (USDOT) to prepare a report that would

    (1) assess the status of dedicated short-range communications [DSRC] technology and applications developed through research and development;
    (2) analyze the known and potential gaps in short-range communications technology and applications;
    (3) define a recommended implementation path for dedicated short-range communications technology and applications that
        a) is based on the assessment described in paragraph 1; and
        b) takes into account the analysis described in paragraph 2;
    (4) include guidance on the relationship of the proposed deployment of dedicated short-range communications to the National ITS [Intelligent Transportation Systems] Architecture and ITS Standards; and
    (5) ensure competition by not preferencing the use of any particular frequency for vehicle to infrastructure operations.

Congress further directed the Secretary of USDOT to engage the National Research Council (NRC) in an independent peer review of the draft report. This letter report contains the peer review conducted by the committee convened by NRC for this purpose. The names of the committee members and their affiliations are listed in Enclosure A, and biographical information about the members is provided in Enclosure B.

THE NATIONAL ACADEMIES
Advisers to the Nation on Science, Engineering, and Medicine

500 Fifth Street, NW
Washington, DC 20001

Phone: 202 334 2934
Fax: 202 334 2003
www.TRB.org

After the committee was convened, it reviewed USDOT's draft report,[1] referred to here as the "DSRC report." The committee subsequently held one 2-day meeting, at which time it was briefed by the main authors of the DSRC report, USDOT staff, and invited stakeholders; gave the public an opportunity to comment; and deliberated in private on the contents of this report. The names and affiliations of the participants in the meeting are included in Enclosure C. The committee completed its report through a series of conference calls and correspondence. In addition, the committee's report was independently reviewed by external individuals, whose names were not known to the committee at the time of the review. The committee's response to these reviews and subsequent changes to the report were approved by NRC. The names of the reviewers appear in Enclosure D.

EXECUTIVE SUMMARY

In this section the committee summarizes its main conclusions about the DSRC report, which are organized under headings of the charge from Congress.

1. Status of DSRC Technology and Applications

The use of 5.9 gigahertz (GHz) DSRC is appropriate for the connected vehicle initiative. The committee agrees with the DSRC report's arguments concerning the low latency, privacy protection, and other benefits this technology offers compared with other communications technologies for safety-critical messages.

Many applications reliant on DSRC are at an early stage of development and are heavily dependent on what manufacturers choose to implement for vehicle-to-vehicle (V2V) and what infrastructure owners and vehicle and device manufacturers choose to implement for vehicle-to-infrastructure (V2I).[2] Any detailed assessment of the maturity and effectiveness of these applications is premature. The principal references relied on in the DSRC report to estimate the benefits of connected vehicle safety applications are not yet available to the public. Therefore, the committee could not verify them independently. It would be helpful to Congress, program stakeholders, and the public for the National Highway Traffic Safety Administration (NHTSA) to complete its review of these reports expeditiously and make them available by the time the DSRC report to Congress is released.

2. Known and Potential Gaps

With regard to DSRC as the chosen low latency technology for communicating safety-critical information, the committee agrees with the DSRC report conclusion that proposed spectrum sharing in the 5.9 GHz band is the most serious risk and uncertainty for the program, but it is not the only one. The committee believes that unless local area wireless technology (Wi-Fi) and other unlicensed and licensed technologies are determined not to interfere with DSRC, the potential benefits of the program will be severely compromised.

---

[1] *Status of the Dedicated Short-Range Communications Technology and Applications [Draft] Report to Congress.* John A. Volpe National Transportation Systems Center, USDOT, 2014.

[2] V2I is used throughout the report to refer to vehicle-to-infrastructure as well as infrastructure-to-vehicle. This terminology is consistent with the nomenclature used by industry and government.

The committee believes that, contrary to the impression conveyed in the DSRC report, other important unknowns and uncertainties will also need to be resolved for the benefits of the connected vehicle initiative to be realized. As discussed in greater detail in a subsequent section of this report, most of these unknowns and uncertainties depend on how the program will be implemented by government and industry, and many implementation details have not been resolved. On the basis of the questions posed for comment in its August 20, 2014, Advance Notice of Proposed Rulemaking (ANPRM) on Vehicle-to-Vehicle Communications, NHTSA itself apparently realizes the importance of several of these issues.

The following are other unknowns and uncertainties that the report could acknowledge or address in greater detail:

- Spectrum frequency coordination,
- Scalability of DSRC beyond levels tested to date,
- Security and privacy considerations,
- Estimated effectiveness and safety benefits of applications,
- Human factors issues associated with implementation of the applications,
- Certification processes for V2V and V2I equipment,
- Reliability of sensors and related electronic systems on vehicles,
- Priority for safety-critical messages to drivers in complex situations,
- Availability of funding for necessary infrastructure,
- Stability of standards, and
- Liability issues associated with system failure.

These issues are addressed in the section headed "Known and Potential Gaps" in the body of the committee report.

3. Implementation Path

The DSRC report explicitly offers a "vision" of how technologies will be implemented in the future rather than a detailed path for implementation. Since the government may choose to mandate installation of DSRC in new vehicles but not mandate applications for the foreseeable future, the committee agrees that stating a vision is more appropriate at this time than defining a detailed implementation path. However, several of the unknowns and uncertainties mentioned above (e.g., certification) are on the critical path to implementation and deserve greater discussion in the draft. In addition, while USDOT has announced a technical plan to build and test the security system, which is critical in enabling any V2V or V2I implementations, the plan to address the policy issues related to that system, such as funding and operation, appears to be unresolved.

4. Consistency with ITS Architecture and Standards

The report includes appropriate guidance on the relationship of the proposed deployment of DSRC to the National ITS Architecture and ITS Standards.

5. Preferencing of Technologies for V2I Applications

The report appropriately preferences 5.9 GHz DSRC for crash-imminent safety applications but allows for other communication technologies for other applications.

In summary, strengthening of the DSRC report in a variety of ways, which are itemized in the following sections, would be beneficial.

COMMITTEE REPORT

Introductory Comments

The connected vehicle initiative is a broad government–industry effort to rely on communications among vehicles (V2V) and between vehicles and infrastructure (V2I) that could provide major improvements in motor vehicle safety and offer many services to travelers. Research, concept development, prototype development, and testing have been taking place for more than a decade. In 1999, the Federal Communications Commission (FCC) allocated 75 megahertz (MHz) of spectrum in the 5.9 GHz DSRC band for use in ITS vehicle safety and mobility applications.[3] On August 20, 2014, NHTSA issued an ANPRM concerning a Federal Motor Vehicle Safety Standard (FMVSS) on V2V communications (Docket No. NHTSA-2014-0022). In the ANPRM, NHTSA indicates that its interest is in mandating the installation of DSRC in future vehicles and that it is not proposing the mandating of any applications based on DSRC at this time.

In reviewing the draft DSRC report, the committee interpreted Congress's intent in Item 5 of the charge—to "ensure competition by not preferencing the use of any particular frequency"—to refer to "communications technology" rather than the radio frequency per se. USDOT should clarify its interpretation of this term in its report. In addition, in reviewing potential connected vehicle applications, the committee focused on safety applications, as did USDOT's DSRC report itself, with the exception of Item 5 of the charge, which includes communications technologies for V2I safety and nonsafety applications.

The committee's responsibility was to review the DSRC report, which relies heavily on other documents for substantiation of claims that it makes. A main reference relied on is *Vehicle to Vehicle Communications: Readiness of V2V Technology for Application,*[4] which is referred to here as the "Readiness report." That report is the principal technical reference accompanying NHTSA's 2014 ANPRM on vehicle-to-vehicle communications. The committee did not review the Readiness report or any of the other references in the same depth as it did the DSRC report. The DSRC report references research conducted for NHTSA by the Crash Avoidance Metrics Partnership (CAMP).[5] The committee was briefed on the results of some of the CAMP research,

---

[3] FCC, 47 CFR Parts 2 and 90, November 26, 1999. http://www.gpo.gov/fdsys/pkg/FR-1999-11-26/html/99-30591.htm.

[4] Report No. DOT HS-812-014. NHTSA, USDOT, August 2014.

[5] CAMP, formed in 1995 by Ford Motor Company and General Motors Corporation "to accelerate the implementation of crash avoidance countermeasures in passenger cars to improve traffic safety," is a research organization that includes employees of up to nine automotive companies. It has a cooperative agreement with USDOT to conduct research on new and emerging technologies for vehicles. Because the costs are shared, both

but it did not have access to the reports themselves, which are still under review by NHTSA.

Before the substance of the DSRC report itself is discussed, a note about terminology is needed. The term "DSRC" in the report is used to refer to different aspects of DSRC. For example, the DSRC report in 2.II.A refers to "(1) DSRC's low latency and high availability, and (2) the ability to provide security, privacy, and a no-subscription fee policy." Here, aspects of DSRC spectrum allocation, communications protocols, software, and system design are all mixed in one statement. The DSRC report would provide the reader a better understanding if it enumerated the various aspects of DSRC, indicated what entities [e.g., FCC, USDOT, state departments of transportation, original equipment manufacturers (OEMs)] influence each, and clarified which aspects are indicated when the term "DSRC" is invoked.

A key metric invoked in discussing DSRC's suitability for crash avoidance applications is "latency." The USDOT report makes the point that DSRC technology is uniquely capable of meeting the "low latency" requirements of safety applications; however, the report uses multiple definitions of latency, spanning at least three orders of magnitude. Furthermore, the definitions are used in contradictory comparisons with other technologies, and it is often not possible to infer which of them is being invoked.

For example, in Section 2.II.A of the DSRC report (page 20), under the heading "Speed of Transmission," the report states that DSRC has latency "well under 100 microseconds." On the same page the report notes two definitions of latency. First, "the lower limit of latency is determined by the physics of the medium." Second, "latency further includes any delays in the transmission processing, propagation delays through the medium ... and receiving processing." Does the "well under 100 microsecond" latency claim on that page include those processing delays? It does not appear to, and it does not even appear to be consistent with the "physics of the medium." Given that basic safety messages (BSMs) are on the order of 3 kilobits and are transmitted at 6 megabits per second, the transmission time from first bit to last bit is on the order of 500 microseconds.

Later in Section 2.II.A (page 22), the DSRC report notes that some technologies impose a delay to "join the network" but that "latency estimates assume that the technologies are already part of a network." This assumption does not appear to be reflected consistently in the technology comparisons reported in Tables B-1 and D-1 of the report. Table B-1 shows DSRC latency in the range of 0.2 to 15 milliseconds (msec). Table D-1 uses the 0.2 msec value for DSRC. However, both tables report Wi-Fi latencies in the range of 3 to 5 seconds, which can only be accurate if the time to join the network is included. The tables are inconsistent with respect to the latency they attribute to LTE (3G) cellular, with Table B-1 reporting 79 to 100 msec and Table D-1 reporting 1.5 to 3.5 seconds. These tables apparently apply different definitions of latency. The claims of sub-millisecond latency for DSRC in Tables B-1 and D-1 also appear to neglect the channel access delay that is fundamental to the IEEE 802.11 MAC protocol. In a dense vehicle environment, a given sender can expect to wait up to 10 msec before beginning transmission.[6]

Finally, the DSRC report invokes a third definition of latency in Appendices B and D. The third definition includes not only physics and processing but also message scheduling delay. Appendix B.II.B (page 79) states that "the latency will be driven by the message repeat interval

---

parties need to agree before research conclusions are published. Recently, CAMP has been performing research on spectrum issues and has provided those results to NHTSA in the form of reports.

[6] IEEE 802.11-2012 Wireless Local Area Networking standard. See http://standards.ieee.org/findstds/standard/802-11-2012.html.

and the communications latency." The default BSM repeat interval is 100 msec (for example in the Safety Pilot Model Deployment).[7] Similarly, Appendix D of the report cites the NHTSA Vehicle Safety Communications (VSC) report (footnote 140),[8] which is the source of Table D-1 and which defines "allowable latency" as "the maximum duration of time allowable between when information is available to be transmitted and when it is received (e.g. 100 msec)." Thus, the report variously claims that DSRC can achieve BSM latency on the order of hundreds of microseconds to hundreds of milliseconds. It also contains contradictory claims about whether "network join" latency is included or excluded in technology comparisons. Collectively these inconsistencies obscure the key point that DSRC does have latency advantages over other technologies, particularly those requiring multiple hops between source and destination and those requiring "network join" delays. If USDOT wishes to use multiple definitions of latency, which one is being invoked in any given instance should be clear, either explicitly or implicitly from the context.

1. Status of DSRC Technology and Applications

The congressional charge to USDOT to assess the readiness of 5.9 GHz DSRC for the connected vehicle initiative expresses equal interest in the status of the applications that will rely on the communications technology. As mentioned, the NHTSA ANPRM indicates that NHTSA would only mandate that future vehicles be equipped with DSRC devices. NHTSA does not propose to mandate any applications at this time. Instead, NHTSA will rely on industry to develop the applications according to performance metrics that NHTSA might specify in future FMVSSs or in the New Car Assessment Program (NCAP) rating criteria.

The committee views fundamental DSRC technology to be adequately tested and appropriate as the preferred communications medium for low latency connected vehicle safety applications, but it notes that some related upper protocol layers, and particularly applications, are less mature. The committee agrees with the DSRC report's assessment of its advantages over other technologies for safety-critical communication.

The DSRC report appropriately describes a handful of crash-imminent safety applications as having been the subjects of intensive work by CAMP and individual OEMs. Other safety warning and nonsafety applications are more conceptual at this stage and are described appropriately in the report. As noted earlier, many of the reports from CAMP projects are not yet publicly available. Therefore, it is critical that NHTSA complete its reviews and release them, preferably by the time the DSRC report is finalized this summer, so that they will be available to Congress, program stakeholders, and the public.

2. Known and Potential Gaps

The draft DSRC report to Congress, Section 3.IV.C.1, indicates that spectrum sharing with unlicensed users is "the one known [technical] gap" in the performance of V2V communications.

---

[7] Vehicle Awareness Device Specification, USDOT, Document USDOTVAD, Version 3.5, Walton Fehr, December 2011.

[8] NHTSA, Vehicle Safety Communications Project Task 3 Final Report: Identify Intelligent Vehicle Safety Applications Enabled by DSRC. DOT HS 809 859. (March 2005, p. 46).

The report notes that administration policy has directed "the FCC to identify and make available 500 Megahertz (MHz) of spectrum over the next 10 years to share with wireless broadband use." It also notes that a group of federal agencies advising FCC has recommended allowing the sharing of spectrum that overlaps with the 5.9 GHz band and that federal legislation directs the Department of Commerce to evaluate "spectrum-sharing technologies and the risk to users" in this same band. Congress has also asked FCC to consider allowing unlicensed WiFi operations in this band as well, which if decided would likely impact the ability to utilize the frequency for connected vehicles. Such a decision has not yet been made.

In a similar manner, the International Telecommunications World Radio Conference 2015 is considering proposals to expand the band above 5,925 MHz for sharing with broadband systems such as Wi-Fi. The DSRC report concludes that wireless industry spectrum-sharing proposals made to date fall short of ensuring that unlicensed Wi-Fi would not interfere with V2V communications and states that USDOT has research under way evaluating possible coexistence with unlicensed users in the 5.9 GHz band. The committee agrees that resolving the interference issue from potential unlicensed as well as licensed broadband uses in the band used by DSRC is critically important.[9]

However, the DSRC report is silent about other possible sources of interference from other unlicensed as well as licensed devices. The committee notes that there are regions of the country, including the greater Washington, D.C., area, where military radars have precedence in or near the DSRC band could interfere with V2V and V2I safety messages. DSRC will not receive protection within 75 kilometers of radars in locations specified in §47 CFR 90.371(a). Because the vehicle safety messages will be broadcast frequently, intermittent military radars may not pose a conflict, but this issue should be checked. In addition, §47 CFR 18.305 permits industrial, scientific, and medical (ISM) equipment to operate in the band 5,800 MHz ± 75.0 MHz, which encompasses Channel 172, and to transmit unlimited radiated energy, and radio-frequency identification systems typically operate in ISM and unlicensed bands, including 5,725 to 5,875 MHz. Further study is needed to determine the interference potential of these uses and, depending on severity, means for ensuring reliable reception of critical safety messages.[10] FCC rules for international intercontinental fixed satellite systems are codified in 47 CFR 2.108, which requires case-by-case electromagnetic compatibility analysis that demonstrates compatible operations with all users of this band including DSRC. It is the committee's understanding that this coordination with DSRC has been performed.

In discussions with the committee, authors of the DSRC report and USDOT staff with expertise in spectrum-sharing issues indicated that USDOT interaction with FCC on spectrum sharing is mediated through the National Telecommunications and Information Administration (NTIA) of the Department of Commerce. Committee members with expertise in this area and familiar with FCC operations were unaware that USDOT technical staff are not working directly

---

[9] The report does not address whether USDOT is participating in appropriate ITU-R World Radio Conference preparation, the NTIA–FCC broadband spectrum Policy and Plans Steering Group, and related activities to ensure that the spectrum requirements of DSRC are considered in any reallocation of spectrum.

[10] This study need not result in moving BSM Channel 172 to a different frequency. For example, a second DSRC BSM channel, supplemental to but not adjacent to Channel 172, could be considered to avoid the possibility of interference. This is similar to what was successfully adopted by the Automatic Identification System (AIS) used in maritime service for the safety of navigation [(ITU-R) Recommendation M.1371 (series)]. Use of two BSM channels would require duplication of the on-board equipment receiver/decoding component to monitor two DSRC channels simultaneously.

with their counterparts in FCC, who could be helpful in identifying and addressing some of the possible sources of interference in the 5.9 GHz band.

The committee also believes that V2V and V2I should be considered by FCC as a safety service. FCC Regulation 47 CFR Section 2.1 and International Telecommunication Union Radiocommunication Sector (ITU-R) 2012 Radio Regulations §1.59 both define safety service as "any radiocommunication service used permanently or temporarily for the safeguarding of human life and property." Usage of DSRC Channels 172, 178, and 184 appears to meet this definition, and planned uses of the remaining DSRC channels are also expected to meet this definition. FCC designation of 5.9 GHz DSRC as a safety service would afford the DSRC (V2V and V2I) safety services a higher level of protection from interference on the 5.9 GHz band.[11] Greater proactivity on the part of USDOT staff with regard to threats to connected vehicle communications appears to be needed.

Other unknowns and uncertainties could affect DSRC and safety applications, and not all of them are identified or adequately addressed in the DSRC report.

Spectrum Frequency Coordination

In Section 3.IV.C.2, the DSRC report recognizes as a "gap" the lack of an organization to coordinate frequency use, particularly among safety and nonsafety applications. It raises appropriate questions about which organization would fulfill this role, the characteristics it should have, and how its activities would be funded. Other questions *not* raised in the DSRC report are equally important. The organization that eventually becomes the frequency coordinator will need policies or regulations in place to guide its activities and will need enforcement mechanisms. As DSRC radio technology is deployed, numerous unforeseen uses will likely be found for the spectrum, and the frequency coordination effort will need to make clear which ones are allowed and how to prevent the ones that are not allowed. The following are examples of possible uses:

- State department of transportation–developed safety application on the standard DSRC safety channel;
- Commercial safety application, perhaps advertisement or subscription supported, perhaps not using USDOT security certificates, on a DSRC safety or service channel; and
- Commercial nonsafety application (e.g., notification of roadside service ahead) on a service channel.

Acknowledgment by the DSRC report of these issues and how they are being addressed by the connected vehicle initiative would be useful.

---

[11] For example, the ITU-R Radio Regulations state:

4.10 Member States recognize that the safety aspects of radionavigation and other safety services require special measures to ensure their freedom from harmful interference; it is necessary therefore to take this factor into account in the assignment and use of frequencies.

15.37 § 29 An administration receiving a communication to the effect that one of its stations is causing harmful interference to a safety service shall promptly investigate the matter and take any necessary remedial action and respond in a timely manner.

Scalability

The issue of scalability concerns whether DSRC technologies can manage the volume of messages that would be broadcast within the 300-meter range of DSRC devices without being overwhelmed at expected maximum vehicle density in traffic. The potential could be even greater in urban environments with the pedestrian applications being contemplated. In Section 3.IV.B, the DSRC report notes that unpublished NHTSA and CAMP research demonstrates that V2V communications "perform reliably" with up to 200 vehicles and that ongoing research will estimate the number of vehicles at which channel congestion would be significant. Without access to the results of the CAMP research, the committee is not in a position to verify this conclusion. The NHTSA Readiness report (page 109) provides an example indicating that up to 800 vehicles could be within DSRC range on a congested freeway. The Readiness report (page 110) also discusses proposed research to test scalability at maximum traffic volumes and research into possible mitigation measures. The committee agrees that such research is important and that the uncertainty about channel congestion remains a major concern for the V2V initiative until it is resolved through technical analysis and testing.

The DSRC report should recognize the importance of scalability and describe work planned or under way to address the potential for interference at likely maximum levels of safety messaging.

Security and Privacy Considerations

The DSRC report addresses one aspect of security, recognizing that public acceptance of V2V and V2I will depend on having a rigorous process for guaranteeing the credentials of devices broadcasting messages and warnings to ensure accuracy of messages and avoid spoofing. However, discussion of this issue is scattered throughout the DSRC report, and treatment of this topic is less than cohesive. The DSRC report notes on page 43 that "USDOT, CAMP, and security experts have developed an initial security and privacy protection solution that will be prototyped and tested in 2015." The report is silent with regard to unanswered questions about the organization that would perform the credentialing service and how its operations would be funded. The Readiness report, for example on pages 158–207, treats the technical and policy aspects of a security credentialing management system (SCMS) in considerable detail and lays out options for how the system might be organized and managed. NHTSA's ANPRM states that it will solicit statements and comments from organizations that would be interested in carrying out the SCMS management function (https://www.federalregister.gov/articles/2014/10/15/2014-24482/vehicle-to-vehicle-security-credential-management-system-request-for-information; Docket No. NHTSA-2014-0023), indicating the agency's appreciation of the importance of this issue. The committee views the policy questions about SCMS management and funding as central issues that require resolution and more discussion in the DSRC report.

However, credentialing is only one of a number of security issues that will need attention. Work in recent years demonstrating the vulnerability of automobiles to standard hacking techniques, including remote hacking, is illustrative of the challenges faced in securing software-intensive systems. The difficulty is compounded by the longevity of vehicles, which greatly exceeds the lifetime of many other software-intensive systems and complicates such tasks as monitoring the security of devices or ensuring that their software is patched to address security flaws. Security will need to be addressed as a property of the overall system rather than of individual components. For example, unless sensors are properly certified and protected, a credentialed communication device fed by a faked sensor could be removed from a car and used to broadcast spurious warnings.

Operating agencies plan to utilize the latest USDOT-led security system design, which will affect operations and maintenance budgets. The additional operations and maintenance cost will need to be accounted for by any agency deploying connected vehicle systems and technology. However, potential third party business models to support, maintain, and operate (or any combination of these functions) the security system are also being considered.

With respect to privacy, the report focuses primarily on avoiding personal information disclosure and does not, except by reference to the standard IEEE 1609.2-2013 address the issue of linkability (the ability to link identifiers through serial numbers, certificates, or other means). The committee is concerned that the DSRC report does not adequately distinguish between privacy expectations in connection with an authorized service provider and privacy expectations in connection with third parties who may monitor the wireless medium. The user may have different expectations in those two regards. Thus, a subscription model, for example, may be entirely acceptable for some services (e.g., tolling), despite the lack of privacy the user obtains with respect to the provider. Statements like "non-DSRC communications companies will need to find a business model that is not reliant upon knowing information about the user" (page 28) do not seem to be well supported in relation to nonmandatory, "opt-in" types of applications. Possible techniques for subscription service privacy, such as cryptographic schemes akin to those developed for digital cash and anonymous credentialing systems, were not mentioned.

The committee is also concerned that USDOT may expect a uniform security solution to be applied to all communications between DSRC devices. The committee believes that DSRC use cases will arise that do not necessarily rely on the security and privacy mechanisms required for mandatory use cases, such as V2V safety. The report should provide a clearer explanation of applications that require security and privacy protection, applications that do not and why they do not, and which technologies and frequencies the latter applications would be expected to use.

Estimated Benefits

The committee recognizes the difficulty of precisely projecting the benefits of the connected vehicle initiative at this point. Doing so with high reliability would require a large-scale test in a naturalistic driving environment that would be more complex, extensive, and expensive than the field operational test involving 2,800 vehicles conducted previously in Ann Arbor, Michigan. Thus, the DSRC report appears to be justifiably cautious in its estimates of benefits, which the committee views as another source of uncertainty.

However, the benefits estimated on pages 30 and 31 (Section 2.II.C) of the DSRC report are stated with high precision and without equivocation, even though NHTSA's own ANPRM describes the same estimates as "very preliminary." The estimates on these pages of the DSRC report, as presented, appear to assume widespread deployment and 100 percent effectiveness, with no adjustment for deployment penetration and estimates of application effectiveness. (These estimates are derived from CAMP and other NHTSA research, much of which has not yet been made public.) Estimates of how DSRC applications, benefits, and costs might unfold over time as more and more vehicles are equipped would be much more relevant. In the committee's view, considerably more research, and independent review of that research, appears to be needed to estimate more credibly the effectiveness of connected vehicle applications.

The DSRC report, pages 30 and 31 (Section 2.II.C), cites benefit estimates derived from two V2V applications [intersection movement assist (IMA) and left-turn assist (LTA)] that would improve intersection safety. It fails to describe the benefits that are likely to come from V2I intersection safety applications and how these benefits have been considered in ascribing safety benefits to the IMA and LTA V2V applications. If these benefits were not considered in ascribing safety benefits to the IMA and LTA V2V applications, this omission should be made explicit in the report along with the rationale for the omission.

If they are widely deployed, the V2I intersection safety applications appear likely to produce more intersection safety benefits during the initial decade of DSRC deployment than the V2V intersection safety applications. For the long ramp-up period until a majority of vehicles are DSRC equipped, V2V applications will not be able to sense vehicles that are not DSRC-equipped. Infrastructure-based traffic detectors can sense non-DSRC equipped vehicles and can provide the information about those vehicles to equipped vehicles by using I2V communication. In the long run, most vehicles may have DSRC radios, but many years could elapse before the number of equipped vehicles surpasses the number of nonequipped vehicles, since 20 years or more may be needed for the full vehicle fleet to turn over. A clear description of how this situation has been addressed in arriving at the benefits estimation for the IMA and LTA applications would be desirable, as well as an explanation of why benefits estimates have not been included for the additional V2V safety applications that have been developed and tested under USDOT cooperative research with CAMP.

Human Factors

One important issue, about which the report is silent, is how drivers might adapt to V2V and V2I warnings and whether they will become less attentive to the driving task because of overreliance on technology. Another human factors challenge could arise if applications are not standardized, which could result in greater risks for drivers using unfamiliar vehicles. The obvious tension is between OEM innovation and competitive differentiation on the one hand and greater uniformity on the other. For example, standardization of such applications could result in a "least common denominator" implementation approach, since such a standard would necessarily be based on the capabilities of the least capable vehicle, unless a mandate required that the enhanced sensor, onboard processing, and driver–vehicle interface capabilities of more advanced vehicle platforms be included on all cars. However, such a requirement would likely increase the base price to the consumer of the least expensive new vehicles. NHTSA may have research planned or under way that addresses human–systems integration (HSI) issues, but such research is not mentioned in the DSRC report. In contrast, the NHTSA ANPRM acknowledges the importance of human factors research and seeks public comment on how it might address connected vehicle HSI concerns through such research.

Certification Processes

In Section 3.IV.C.3, the DSRC report lists as an "open item" certification test procedures that will be needed to ensure that vehicular and infrastructure devices perform as intended. It notes that some certification procedures exist in draft form and refers to Appendix G of the report for additional details. However, Appendix G is incomplete; it merely describes the layers of performance that will need to be certified without providing details about the procedures for doing so.

The DSRC report focuses on DSRC device certification, but performance of other system components is not addressed. V2I efficacy assumes accuracy of signal phase and timing (SPaT) information provided to vehicles, but there is no known certification program guaranteeing that traffic controllers generate accurate SPaT data. (V2I efficacy at full scale-up also assumes ubiquitous accurate intersection map geometry data, but how these data would be systematically produced across the nation is not indicated in the DSRC report.)

The DSRC report notes that new hardware has to be certified by FCC, but the process has not yet been laid out. FCC has regulatory authority over devices that transmit radio frequencies, which, for example, could provide a means of certifying aftermarket DSRC devices. FCC could, by rulemaking, apply a recognized DSRC equipment performance standard and test procedure in addition to its existing certification requirements, as it has often done for safety equipment used in the maritime service (e.g., 47 CFR §80 Parts 1101–1103).

NHTSA's Readiness report assumes that vehicle OEMs will be responsible for certification and acknowledges (on page xvii) that OEMs and V2V device manufacturers will have a "significant testing obligation" to guarantee interoperability with other devices and security credentialing. The Readiness report notes in several places the proposed NHTSA research projects to address aspects of certification. The committee views certification issues for OEMs and aftermarket devices as more of a challenge than is conveyed in the DSRC report. Although resolution of this challenge will presumably largely be the responsibility of the private sector, it is a fundamentally important step on the path to deployment and therefore should be discussed more prominently in the DSRC report.

In addition to certification, some process will need to be determined for verifying that DSRC radios and devices continue to perform as intended. Ongoing credentialing of devices that might be in the field for 20 years is an important aspect that needs to be considered.

Development of performance and test standards for DSRC systems (including aftermarket systems) by recognized national and international standards organizations, to be incorporated by reference into the certification regulations of an agency (such as FCC) having appropriate statutory authority to certify such equipment to ensure that vehicle devices perform as intended over their lifetime of use, would assist in the certification process. It would also be helpful for USDOT and FCC staff to meet routinely to address DSRC spectrum and certification issues addressed in this review, with participation by NTIA staff where appropriate when national spectrum allocation policy issues are considered, as the U.S. Coast Guard and FCC have done for many years to good effect. Frequency coordination with Canada and Mexico could be included. Congress and program stakeholders would benefit from a discussion in the DSRC report concerning how USDOT staff are interacting with FCC to prepare for performance and test standards for DSRC systems.

Sensor Reliability

Another obligation that will fall on the private sector is to ensure that the vehicle sensors that provide the basic data for warnings are themselves accurate and that the most safety-critical warnings receive priority over others. The DSRC report states on page 3 that "the incorporation of a communications capability within vehicle sensor systems permits data on emerging threats and hazards in the roadway to be gathered from multiple external sources . . . and fused with on-board data." This integration with vehicle sensor systems needs to be carefully understood. How thoroughly V2V data will be integrated with data from on-vehicle sensor and positioning systems initially is not clear; the data may be loosely coupled. As the evolution of external data with onboard sensor data reaches a point of fusion when both are necessary and critical for making judgments about safety situations, the accuracy and robustness of the onboard sensor data need to be assured. A "hard" sensor failure is relatively easy to identify, and appropriate action can be taken to correct the situation via built-in redundancy or some other form of data correction. However, a gradual, long-term degradation of a sensor (such as the sensors and systems described below) resulting in a gradual increase in the inaccuracy of the data provided can be a significant problem, as can intermittent failures. This type of error, especially in the case of a critical data element, can cause significant problems in a connected vehicle scenario when, for example, 200 or more cars in a 300-meter range are all receiving some form of erroneous data from a given sensor. Strict built-in test parameters and limits will need to be defined for data that are derived from sensors, especially those critical to safety applications (e.g., rate sensors, accelerometers, speed sensors, steering angle and torque sensors, antennas, GPS receivers, and visual data). For example, the yaw sensor used for electronic stability control (ESC) application has a continuous built-in test that monitors the performance of the sensor and all of its associated electronics to determine whether the sensor output is within predefined "acceptable limits." If the sensor demonstrates an unacceptable performance, the ESC is turned off within fractions of a second and the vehicle is operated in the manual mode. DSRC systems will need to use position data from navigation receivers such as GPS that have been authenticated and certified to ensure that requirements have been effectively met.

Reliability of sensors may have been studied as part of the field tests in Ann Arbor, Michigan, but the report of these tests has not yet been made public. If a study or report pertaining to ensuring the reliability of the electronics (and associated sensors) does exist, it needs to be referenced in the report and made public for independent investigation and analysis. The DSRC report should discuss this issue and describe how it is being addressed in the connected vehicle initiative.

Safety-Critical Messages and Multicriteria Decision Making

The related issues of safety alert prioritization and driver overload are not addressed in the report. These topics are not limited to V2V and V2I safety systems; they are of concern to all safety warning systems and continue to receive much attention, and thus, should be included in the DSRC report. Hierarchical data aggregation and multicriteria decision-making techniques could be applied in such situations, which must be addressed for the system to work effectively. This issue is expected to be addressed in a proprietary manner by individual OEMs with regard to their advanced safety systems features. Whether this would be an appropriate area for industry standards or federal regulations, or both, is not clear due to the diversity in applications and technical capabilities of various vehicle platforms. However, there has been some previous informative research on this subject.[12] Even though this research was not entirely based on DSRC-enabled safety systems, reference to it would strengthen the DSRC report.

Funding for Infrastructure

As USDOT is undoubtedly well aware, funding for the deployment, operation, and maintenance of the roadside hardware and software necessary for V2I communication is unresolved. Congress has struggled to find ways to fund the upkeep of the existing highways and bridges on the

---

[12] For more information on relevant projects, the reader is referred to
http://www.nhtsa.gov/DOT/NHTSA/NRD/Multimedia/PDFs/Crash%20Avoidance/2008/DOT-HS-810-905.pdf;
Human Factors for Connected Vehicles: Effective Warning Interface Research Findings, DOT HS 812 068, September 2014; and *Crash Warning System Interfaces: Human Factors Insights and Lessons Learned*, J. L. Campbell, C. M. Richard, J. L. Brown, and M. McCallum, DOT HS 810 697, January 2007.

federal-aid system, and DSRC will add new, more sophisticated technologies that will require ongoing upkeep expenditures. Moreover, many of the nation's busiest intersections that would be priority candidates for V2I infrastructure and applications may not even be on the federal-aid system, and a new financial burden on county and municipal governments that can barely afford to retime traffic signals on a regular basis would be imposed. Until these issues are addressed, rollout of V2I applications on a broad scale appears questionable, and this should be noted in the DSRC report.

Stability of Standards

The DSRC report notes on page 40 that standards groups are working on "next versions for publication in the near future" and on page 41 notes that they are "developing additional protocols. . . ." The stability of underlying communications technologies and standards is a known gap that should be acknowledged in the DSRC report.

Liability for System Failures

The DSRC report is silent about liability issues for both the public and the private sectors resulting from failures of the connected vehicle system to provide warnings as intended. In contrast, the Readiness report acknowledges that this is a major concern of some segments of industry and devotes a full chapter to the issues (pages 208–215). The NHTSA position stated in the Readiness report is that the V2V and V2I warnings that drivers will receive are not different in kind from the types of warnings that vehicles already provide to motorists from onboard sensors. According to the Readiness report, manufacturers can manage their liability in this regard by complying with industry standards and federal regulations and by providing adequate guidance to vehicle owners through manuals and other warnings. In addition, liability concerns as related to the public sector will need to be addressed. The committee has no viewpoint to express on the liability issue, but because of the controversy and concern about it in some segments of the industry, the DSRC report should acknowledge the industry concerns and how NHTSA views them as summarized above.

3. Implementation Path

Because the connected vehicle initiative will depend on industry development and deployment of applications, the DSRC report appropriately describes future applications at a fairly conceptual level and describes a vision for how these systems might be deployed in the future, rather than specifying a specific path for implementation. However, as noted above, several critical steps in the implementation path receive limited attention in the DSRC report (frequency coordinator, operation and funding of SCMS, equipment certification, equipment accuracy, infrastructure funding, and resolution of liability concerns). Obviously, additional work is required in these areas for implementation to proceed. Such issues could influence the effectiveness of the applications and the willingness of OEMs to deploy these applications in their vehicles. Therefore, the timeline and implementation milestones in the vision could be less certain than might be implied.

4. Consistency with ITS Architecture and Standards

Section 4.II.B of the DSRC report, Role of the National ITS Architecture in Deployment, provides good guidance on the close relationship of the proposed deployment of DSRC to the National ITS Architecture. The report identifies the purpose and role of the National ITS Architecture for state and local transportation agencies in meeting their "needs in planning for ITS deployments while ensuring nationwide interoperability." The report outlines USDOT's development of the Connected Vehicle Reference Implementation Architecture (CVRIA), initiated in 2012, with the intent of supporting deployment of connected vehicle applications and technologies. The CVRIA currently remains under development and once "sufficiently mature" will become part of the National ITS Architecture. The report also highlights USDOT's work on a stakeholder community software tool to develop connected vehicle architectures, known as "Set-IT." The software has been released in an alpha version and is available to the stakeholder community to support deployment activities. Development of training materials to support the tool's use is also under way.

Section 3.III includes guidance on the relationship of the deployment of DSRC to ITS standards. It explicitly lists the relevant standards developed by the Institute of Electrical and Electronics Engineers (IEEE) and the Society of Automotive Engineers (SAE) with support of the ITS Joint Program Office's Standards Program. Appendix D.II.C.ii provides more details on individual standards.

USDOT recognizes that key standards are in a revision cycle, potentially breaking backwards compatibility and containing new protocols. In the future, standards may be periodically updated and revised to correct problems or add new features, but this could affect the validity of testing results produced to date. The effectiveness of the deployed system may be compromised if it is not built on a stable set of standards. The DSRC report notes on page 40 that standards groups are working on "next versions for publication in the near future" and on page 41 notes that they are "developing additional protocols. . . ." Of less importance is a misstatement on page 39 of the DSRC report: "Wi-Fi standard IEEE 802.11 describes the performance of DSRC." A more accurate statement would be the following: "IEEE standard 802.11 specifies the lower-layer behavior of radios used in DSRC."

5. Preferencing of Technologies

The DSRC report makes a convincing case that DSRC technologies are the best choice for low latency communications. Section 2.II.B expresses USDOT's intent to explore all wireless technologies and summarizes USDOT and other research investigating the range of options. The section notes that the connected vehicle initiative can rely on technologies other than DSRC (potentially operating at non-DSRC frequencies) for non–low latency V2I communications. However, the concluding page of the section (page 28) notes that any non-DSRC technologies participating in safety messaging, presumably even those that are not crash imminent (such as an indication that there is fog 1 to 2 miles ahead), would face major challenges in maintaining the necessary level of security, in demonstrating that they can maintain a sufficiently high level of reliability, and in building a business model that is not dependent on information concerning the user.

Conclusion

In conclusion, the committee, as charged, reviewed the USDOT *Status of the Dedicated Short Range Communications Technology and Applications [Draft] Report to Congress.* The committee's findings are summarized in the Executive Summary and are not repeated here. Specific guidance on strengthening the DOT draft report is provided in the body of this report. On behalf of the committee, I express our appreciation for the opportunity to be of service in furthering the important and valuable connected vehicle initiative. We hope that you find our comments on the draft DSRC report to Congress to be constructive in this regard.

Sincerely,

Dennis Wilkie (NAE)
Committee Chair

cc: Dale Thompson, USDOT, ITS-JPO

Enclosure A Committee Roster
Enclosure B Committee Biographical Information
Enclosure C Meeting Participants
Enclosure D Acknowledgment of Reviewers

Enclosure A

**Committee to Review the USDOT Report on
Connected Vehicle Initiative Communications Systems Deployment**

Dennis F. Wilkie, NAE, Ford Motor Company and Motorola, Inc. (retired), Bonita Springs, Florida, *Chair*
David E. Borth, NAE, consultant, Palatine, Illinois
Socorro (Coco) Briseno, California Department of Transportation, Sacramento
Collin L. Castle, Michigan Department of Transportation, Lansing
Joseph D. Hersey, Jr., JoeCel Engineering and Consulting, LLC, Colesville, Maryland
John B. Kenney, Toyota InfoTechnology Center, Mountain View, California
Asad M. Madni, NAE, BEI Technologies, Inc. (retired), Los Angeles, California
John T. Moring, Moring Consulting, Encinitas, California
Tom L. Schaffnit, A2 Technology Management, LLC, Ann Arbor, Michigan
Steven E. Shladover, University of California Partners for Advanced Transportation Technology Program, Richmond

Staff

Beverly Huey, Study Director, TRB
Stephen Godwin, Division Director, Studies and Special Programs, TRB
Jon Eisenberg, Director, Computer Science and Telecommunications Board, DEPS
Amelia Mathis, Administrative Assistant, TRB

Committee Biographical Information

Dennis F. Wilkie, NAE, retired as Corporate Vice President and Chief of Staff for the Integrated Electronic Systems Sector at Motorola, Inc., in 2002. He joined Motorola, Inc., in 1996 after retiring from the Ford Motor Company. During the years at Motorola he was involved in automotive electronic systems, energy systems, and embedded electronic control systems management. He retired from the Ford Motor Company in 1996 as Corporate Vice President for Business Development. He worked at Ford for 28 years, and his work involved the application of control theory and systems engineering to automobiles and the field of transportation. He worked on automotive electronic systems issues as well as infrastructure issues, such as automated highways, automated transportation systems, and ITS. In recent years, he has focused on the utilization of electronics and wireless technology to bring new levels of convenience, safety, and information to the vehicle. He was elected to the National Academy of Engineering in 2000 and is a Fellow of SAE. He holds a BS and an MS in electrical engineering from Wayne State University, a PhD in electrical engineering from the University of Illinois, and an MS in management (Sloan Fellow) from the Massachusetts Institute of Technology.

David E. Borth, NAE, has been serving as an independent consultant in the areas of wireless technology, advanced signal processing, and spectrum engineering since retiring from Motorola in 2010. He also served as a professor of electrical and computer engineering at the University of Illinois at Chicago from 2012 to 2014. From 1980 to 2012 Dr. Borth was with Motorola in Schaumburg, Illinois, where he held a number of positions ranging from member of the technical staff to Corporate Vice President of all wireless research in the company to Chief Technology Officer of the Government and Public Safety Business Unit. While at Motorola Dr. Borth made significant contributions to numerous wireless technologies, including Motorola's implementations of the GSM, TDMA, and CDMA digital cellular systems as well as leading wireless research work focusing on the development of key technologies for broadband wireless systems, including 802.16e/WiMAX, LTE, and 4G systems. He also worked on a variety of emerging wireless technologies, including software-defined radio and cognitive radio. Dr. Borth served as a member of the FCC's Technological Advisory Council (TAC) and of the U.S. Department of Commerce Spectrum Management Advisory Committee (CSMAC) for seven years. He has been issued 31 patents and has authored or co-authored chapters of five books in addition to 25 publications. He is a Fellow of IEEE and a licensed professional engineer in the State of Illinois. Previously, he was a member of the technical staff of the systems division of Watkins–Johnson Company and an assistant professor in the School of Electrical Engineering, Georgia Institute of Technology. Dr. Borth was a member of the Computer Science and Telecommunications Board (CSTB) from 2000 to 2003. He also served on the CSTB committee that produced the report *Information Technology for Counterterrorism: Immediate Action and Future Possibilities*. He received his BS, MS, and PhD in electrical engineering from the University of Illinois at Urbana–Champaign.

Socorro (Coco) Briseno started her career at the California Department of Transportation (Caltrans) in 1990 and is currently Chief of the Division of Research, Innovation, and System Information. This position includes responsibility for the Caltrans research program as well as management of the transportation information required to support the state's public road system decision-making needs. During her time with Caltrans, Ms. Briseno has served in many capacities. She has been Chief of Staff and has worked at the City of West Sacramento, the California Transportation Commission, and the Divisions of Traffic Operations and Planning. Previous positions at Caltrans include Associate Transportation Planner, Rail Transportation Associate, Senior Transportation Planner, and Supervising Transportation Planner. She graduated with a BA in organizational leadership from Chapman University in 2000.

Collin L. Castle, a professional engineer, has worked in the Michigan Department of Transportation (DOT) ITS Program Office for the past 8 years. He serves as the Connected Vehicle Technical Manager with the Michigan DOT, focusing on initiatives related to connected vehicle infrastructure design and deployment and connected vehicle data use for agency applications. This includes analysis of the impacts of connected vehicle infrastructure and applications on safety, mobility, and the environment. He is managing a number of Michigan DOT initiatives, including the Truck Parking Information and Management System (TPIMS), the Connected Vehicle Data Use Analysis and Processing Project, the Vehicle Based Information Data Acquisition System, Cost and Benefit Analysis of Michigan DOT ITS Deployments, and the Weather Responsive Traveler Information System. During his first 5 years at Michigan DOT, he was involved in statewide regional ITS architecture development and conformance, ITS specifications and design standards, construction plan review and approval, ITS laboratory testing, and project development. In a number of these initiatives he served in a project management role, including the Michigan DOT North and Superior Region Road Weather Information System Design–Build project, Statewide Real-Time Probe Data, and Statewide ITS Program Office support contracts. He assists with coordination of research efforts related to automated vehicles and their impact on a road operating agency. This includes support of research and testing efforts by multiple universities, consultants, and industry on the needs of automated vehicles from a traditional transportation infrastructure and technology perspective. He has served in a stakeholder advisory capacity on a number of initiatives and research activities. Among them are the Federal Highway Administration Weather Data Environment, Guidelines for Evaluating the Accuracy of Travel Time and Speed Data—Pooled Fund Study [TPF-5(200)], and the American Association of State Highway and Transportation Officials (AASHTO) Connected Vehicle Footprint Analysis. He was involved in planning efforts for the 2014 ITS World Congress held in Detroit, Michigan, including the Traffic Management Center of the Future and the Belle Isle Technology Demonstrations. Mr. Castle is a graduate of Michigan State University with a BS in civil engineering with a focus on transportation and is a registered professional engineer in the state of Michigan. He is a recipient of the 2014 ITS World Congress—Best Paper Award (Americas) for the I-94 TPIMS project, and the State of Michigan 2014 Good Government Symbol of Excellence and Leadership Coin for leadership in developing innovations in the field of ITS and connected vehicle research. He was a nominee for the 2013 AASHTO Transportation Vanguard Award.

Joseph D. Hersey, Jr., has a consultancy, JoeCel Engineering and Consulting, LLC, that provides contract engineering support work in telecommunications and navigation standards development and review, radio-frequency spectrum management, and spectrum studies. He is a professional engineer with a focus in radio spectrum and maritime telecommunications, and he serves as Secretary for the U.S. National Committee Technical Advisory Group to International

Electrotechnical Commission (IEC) Technical Committee 80, Maritime Radiocommunications and Navigation Equipment. Mr. Hersey served in the U.S. Coast Guard (USCG) from 1975 through 2013 in positions of increasing responsibility. He was chief of the Spectrum Management and Telecommunications Policy Division of USCG, USCG representative to NTIA's spectrum broadband reallocation Policy and Plans Steering Group, agency vice chairman of NTIA's Interdepartment Radio Advisory Committee (which assists the Assistant Secretary in assigning frequencies to U.S. government radio stations and in developing and executing policies), and member of the international technical team that developed the shipborne automatic identification system in widespread use and documented in International Telecommunication Union (ITU) and IEC technical standards. He managed a cutter radar installation team; specified and procured small boat radar; developed and coordinated numerous technical input papers to and represented the United States on International Maritime Organization (IMO) communications subcommittees, ITU Study Groups, and World Radio Conferences; developed and implemented the Global Maritime Distress and Safety System through IMO and ITU and within the United States and developed its initial modernization; authored radar propagation studies; and developed dGPS transmitter frequency selection software based on ground-wave propagation prediction and receiver frequency dependent rejection. He had previously provided USDOT with input on spectrum operations. Mr. Hersey earned an MS and a BS in engineering from Brown University.

John B. Kenney, principal researcher at the Toyota InfoTechnology Center USA, is the lead for the vehicle communication research group. He has recently participated in the V2V-Interoperability (2010–present), V2V-Communication Security (2010–2012), and VSC-Applications (2007–2009) projects, and he represents Toyota in organizations responsible for specifying DSRC standards, including the IEEE 802.11 Working Group (WG), the IEEE 1609 WG, the SAE DSRC Technical Committee (TC), and the European Telecommunications Standards Institute (ETSI) TC ITS. He also serves as a liaison between the IEEE 1609 WG and the VSC consortium. He has served as Secretary of the SAE DSRC TC since 2010 and serves on the ETSI TC ITS Specialist Task Force on Cross-Layer Decentralized Congestion Control (STF469, August 2013–August 2015). On behalf of the VSC consortium, he drafted an analysis of the relevance of the IEEE standards (802.11p and 1609.x) for USDOT's consideration in assessing a potential V2V rulemaking. Dr. Kenney has testified before the U.S. House of Representatives Energy and Commerce Committee's Subcommittee on Communication Technology (November 2013), and he has provided spectrum sharing briefings to members of Congress and their staffs, the staffs of the FCC commissioners, the FCC Office of Engineering Technology, and the White House Office of Science and Technology Policy. He has made several presentations in the IEEE 802.11 DSRC Co-Existence Tiger Team, received a Best Paper Award at the 2013 IEEE International Wireless Vehicle Communication Symposium; and cochaired the IEEE SmartVehicles 2014 workshop and the ACM Vehicular Inter-Networking workshops in 2011 and 2012. His research interests include channel congestion control, spectrum sharing, and wireless communication performance. He earned a PhD and a BS in electrical engineering from the University of Notre Dame and an MS in electrical engineering from Stanford University.

Asad M. Madni, NAE, served as President, Chief Operating Officer, and Chief Technology Officer of BEI Technologies Inc., headquartered in Sylmar, California, from 1992 until his retirement in 2006. He led the development and commercialization of intelligent microsensors and systems for aerospace, military, commercial, and transportation industries, including the Extremely Slow Motion Servo Control System for the Hubble Space Telescope's Star Selector System and the revolutionary Quartz MEMS GyroChip technology, which is used worldwide for electronic stability control and rollover protection in passenger vehicles. Before joining BEI he was with Systron Donner Corporation (a Thorn EMI Company) for 18 years in senior technical and executive positions, eventually as Chairman, President, and CEO. There, he made seminal contributions in the development of radio-frequency and microwave systems and instrumentation, which significantly enhanced the combat readiness of the U.S. Navy (and its allies) and which provided the Department of Defense with the ability (not possible with prior art) to simulate more threat-representative electronic countermeasures environments for current and future advanced warfare training. Dr. Madni is currently an independent consultant; Distinguished Adjunct Professor and Distinguished Scientist of Electrical Engineering at the University of California at Los Angeles (UCLA); and Executive Managing Director and Chief Technical Officer of Crocker Capital, a San Francisco–based private venture firm specializing in emerging technologies. He is an internationally recognized authority with more than 40 years of experience in the design and commercialization of "intelligent" sensors, systems, and instrumentation and signal processing. He also serves as Distinguished Professor at Technical Career Institutes (TCI) College of Technology (the first such appointment in the history of the institute since its founding in 1909 by Nobel Laureate Guglielmo Marconi); as Adjunct Professor in the Computer Science Department at Ryerson University; on advisory boards at UCLA, the University of Southern California, the University of Texas at San Antonio, California State University at Northridge, TCI, *IEEE Systems Journal*, and *AutoSoft Journal*; and as Honorary Editor of the *International Journal on Smart Sensing and Intelligent Systems*. Dr. Madni is the recipient of numerous national and international awards and honors, including the 2014 Tau Beta Pi Distinguished Alumnus Award, the World Automation Congress 2014 (inaugural) Medal of Honor, the UCLA Electrical Engineering 2013 (inaugural) Distinguished Alumni Award, the 2013 UCLA Electrical Engineering Distinguished Service Award, the 2012 IEEE Aerospace and Electronic Systems Society's Pioneer Award, the 2010 IEEE Instrumentation and Measurement Society's Career Excellence Award, the 2010 UCLA Engineering Lifetime Contribution Award, TCI College of Technology's Marconi Medal (the institute's highest honor) and Citation of Honor, the 2008 IEEE Region 6 Outstanding Engineer Award, the 2008 IEEE Region 6 South Outstanding Engineer and Outstanding Leadership and Professional Service Awards, the 2008 UCLA Engineering Distinguished Service Award, the 2006 World Automation Congress Lifetime Achievement Award, the 2005 IEE Achievement Medal, the 2004 UCLA Engineering Alumnus of the Year Award (highest honor granted by the school), the 2004 Distinguished Engineering Achievement Award from the Engineers' Council, the 2003 George Washington Engineer of the Year Award from the Los Angeles Council of Engineers and Scientists, the 2002 UCLA Professional Achievement Award Medal, the IEEE Third Millennium Medal, the Joseph F. Engelberger Best Paper Award at the 2000 World Automation Congress, the California Coast University (CCU) Distinguished Alumni Award (highest honor granted by the university), and the Association of Old Crows Gold Certificate of Merit. He is listed in all major Who's Who publications, including Who's Who in America, Who's Who in the World, Who's Who in Science and Engineering, Who's Who in Technology, Who's Who in Finance and Industry, the International Who's Who of Intellectuals, 2000 Outstanding Intellectuals of the 20th and 21st Centuries, Outstanding People of the 21st Century, and Asian Men and Women of Achievement.

He has been a featured guest on numerous television shows, including CNN with Casey Wyan and BizNews 1 (now CNBC) with Mike Russell. He is a Chartered Engineer, Honorary Professor at the Technical University of Crete and University of Waikato New Zealand, Life Fellow of the IEEE, Fellow of the Institution of Electrical Engineers (United Kingdom), Fellow of the Institution of Engineering and Technology (United Kingdom), Fellow of the Institute for the Advancement of Engineering, Fellow of the New York Academy of Sciences, Fellow of the American Association for the Advancement of Science, Fellow of SAE, Lifetime Fellow of the American Institute of Aeronautics and Astronautics, and Life Fellow of the International Biographical Association. He is an Eminent Engineer of Tau Beta Pi, the National Engineering Honor Society; Honorary Member of Upsilon Pi Epsilon, Honor Society for the Computing and Information Disciplines; Member of Eta Kappa Nu, the Electrical and Computer Engineering Honor Society; Member of Tau Alpha Pi, the National Engineering Technology Honor Society; Honorary Member of Phi Kappa Phi, the Interdisciplinary Honor Society; Honorary Member of Golden Key, the International Honor Society; Member of Sigma Xi, the Scientific Research Honor Society; Member of Delta Epsilon Tau, the International Honor Society; Life Member of the Association of Old Crows; Member of the Internet Society; and Member of the Order of the Engineer. He received an AAS from RCA Institutes, a BS and an MS from UCLA, a PhD from CCU, a DSc (honoris causa) from Ryerson University, a DEng (honoris causa) from Technical University of Crete, and an ScD (honoris causa) from California State University and California State University, Northridge. He is a graduate of the Engineering Management Program at the California Institute of Technology, the Executive Institute at Stanford University, and the Program for Senior Executives at the Massachusetts Institute of Technology Sloan School of Management.

John T. Moring, a consultant, is a systems engineer with extensive experience in developing advanced communications systems, from integrated circuits to international networks. Since 2004 Moring has contributed to the IEEE 1609 standards for DSRC/Wireless Access in Vehicular Environments. For the past 6 years he has served as Cochair of the Working Group and editor of the primary networking and architecture standards (e.g., IEEE Standards 1609.3, 1609.0). In the late 1990s when he founded his consultancy, one of his first projects was to support the new FCC mandate to locate cellular 9-1-1 callers. He designed and ran some of the first field tests of this new feature for major cellular carriers. He was a consultant for the Bluetooth organization for 8 years in the area of certification, as the technology moved from 0 percent penetration to a "must-have" in cellular phones. His clients have included wireless equipment manufacturers and service providers, government entities, universities, think tanks, law firms, and garage-shop startups. Before beginning his consultancy in 1997, he worked for Pacific Communication Sciences, the company that developed the first commercial Internet phone, and he contributed to the associated base station design and development. In the early 1990s he contributed to advanced satellite networking products at Linkabit; in the late 1980s he worked on Internet deployments for the military while at TRW; in the 1980s he helped develop spread spectrum military radios, the forerunner of today's cellular technologies, for Hughes Aircraft. Mr. Moring is named inventor on 13 patents. He earned his MS in electrical engineering from the University of Southern California in 1984 and his BS in electrical engineering from the University of Cincinnati in 1981.

Tom L. Schaffnit, an expert in wireless telecommunications technology with more than 20 years of wide-ranging related experience, is the President of A2 Technology Management, LLC, where he provides strategic technology management support to Honda R&D Americas, Inc., in the area of advanced safety systems enabled by wireless communications. Since 2010, he has also served as President of the Vehicle Infrastructure Integration Consortium. In this role, he provides leadership for a precompetitive consortium of 10 major automakers, focused on development of industry policy positions related to 5.9 GHz DSRC for connected vehicles. In the late 1990s, Mr. Schaffnit was President of CUE Data Corporation, responsible for creating and implementing new ITS datacasting services on a nationwide FM subcarrier data network. Before that he was a senior manager at Deloitte & Touche Consulting Group and a director of telecommunications systems strategies at Nordicity Group, Limited. Mr. Schaffnit earned an MBA in systems management from the University of Manitoba in 1990 and a BSc in industrial management, engineering option, from Purdue University in 1973. He was registered with the Association of Professional Engineers of the Province of Manitoba.

Steven E. Shladover conducts research on automated and connected vehicles systems at the California Partners for Advanced Transportation Technology (PATH) Program at the University of California, Berkeley, a major university research program in ITS, where he has previously served as Deputy Director and Advanced Vehicle Control and Safety Systems (AVCSS) Program Manager. He leads a variety of intelligent transportation research projects at PATH, with an emphasis on connected automation systems to improve mobility. Formerly he was Manager, Transportation Systems Engineering, with Systems Control Technology, Inc. He is active in international standards development, serving as United States Expert and Chairman of the U.S. Working Advisory Group to the International Organization for Standardization's Technical Committee on Intelligent Transport Systems Working Group 14 on Vehicle–Roadway Warning and Control Systems. He was Chairman of the AVCSS Committee of IVHS America/ITS America and currently chairs the Transportation Research Board standing Committee on Vehicle–Highway Automation. He received his SB, SM, and ScD in mechanical engineering from the Massachusetts Institute of Technology.

Meeting Participants

Committee
Dennis F. Wilkie, NAE, *Chair*
David E. Borth, NAE
Socorro (Coco) Briseno
Collin L. Castle
Joseph D. Hersey, Jr.
John B. Kenney
Asad M. Madni, NAE
John T. Moring
Tom L. Schaffnit
Steven E. Shladover

Speakers
Bob Arnold, Federal Highway Administration
Jim Arnold, Office of the Assistant Secretary for Research and Technology
John Augustine, ITS Joint Program Office (ITS-JPO)
Walt Fehr, ITS-JPO
Tim Johnson, NHTSA
Bob Kreeb, NHTSA
Robert Rausch, Transcore, Norcross, Georgia
Mike Shulman, Ford Motor Company, Ann Arbor, Michigan (via conference line)
Steve Sill, ITS-JPO
Dale Thompson, ITS-JPO

Guests
Brian Cronin, USDOT
Dominie Garcia, Booz Allen Hamilton
Dale Kardos, Dale Kardos & Associates, Inc.
Joshua Kolleda, Booz Allen Hamilton
Jade Nobles, Toyota Motor North America, Inc.
Will Otero, Alliance of Automobile Manufacturers
Jack Rickard, AKIN Gump Strauss Hauer & Feld, LLP
Michelle Silva, AKIN Gump Strauss Hauer & Feld, LLP (represented Volkswagen)
Suzanne Sloan, Volpe National Transportation Systems Center

Transportation Research Board Staff
Stephen Godwin
Beverly Huey
Amelia Mathis

Enclosure D

Acknowledgment of Reviewers

This report has been reviewed in draft form by individuals chosen for their diverse perspectives and technical expertise in accordance with procedures approved by NRC's Report Review Committee. The purpose of this independent review is to provide candid and critical comments that will assist the institution in making its published report as sound as possible and to ensure that the report meets institutional standards for objectivity, evidence, and responsiveness to the study charge. The review comments and draft manuscript remain confidential to protect the integrity of the deliberative process. Thanks are extended to the following individuals for their review of this report: Steven M. Bellovin, NAE, Columbia University; David J. Goodman, NAE, New York University; Karl Hedrick, NAE, University of California, Berkeley; Gerald Holzmann, NAE, Jet Propulsion Laboratory; Paul Kolodzy, Kolodzy Consulting, LLC; Bill Legg, Washington State Department of Transportation; Robert Rausch, Transcore; Mike Shulman, Ford Motor Company; and William Whyte, Security Innovation.

Although the reviewers listed above have provided many constructive comments and suggestions, they were not asked to endorse the conclusions or recommendations, nor did they see the final draft of the report before its release. The review of this report was overseen by Maxine Savitz, NAE, Honeywell, Inc. (retired); and Susan Hanson, NAS, Clark University. Appointed by NRC, they were responsible for making certain that an independent examination of this report was carried out in accordance with institutional procedures and that all review comments were carefully considered. Responsibility for the final content of this report rests entirely with the authoring committee and the institution.